The Radio Today guide to the Yaesu FTdx101

By Andrew Barron ZL3DW

Copyright © 2021 by Andrew Barron

All rights reserved.

No part of this book may be reproduced, transmitted, or stored in any form or by any means except for your own personal use, without the express, written permission of the author.

Edition 1.2 January 2024.

ISBN: 9798502302234

This book includes useful tips and tricks for the configuration and operation of the fabulous Yaesu FTDX101D and FTDX101MP transceivers. Rather than duplicate the manuals which describe each button, function, and control, I have used a more functional approach. This is a "how to do it" book with easy-to-follow step-by-step instructions.

The author has no association with Yaesu, any Yaesu reseller, or Yaesu service center. The book is not authorized or endorsed by Yaesu or by any authorized Yaesu dealer or repair center. Research material for the creation of this document has been sourced from a variety of public domain Internet sites and information published by Yaesu including the FTDX101MP and FTDX101D Operation Manual. The author accepts no responsibility for the accuracy of any information presented herein. It is to the best of my knowledge accurate, but no guarantee is given or implied. Use the information contained in this book at your own risk. Errors and Omissions Excepted.

Radio Today is a trademark of the Radio Society of Great Britain www.rsgb.org and is used here with their permission.

Cover graphics by Kevin Williams M6CYB. The transceiver image and Yaesu logo are used with permission from Yaesu (UK) Ltd.

Yaesu, Yaesu Musen, and the Yaesu logo are registered trademarks of Yaesu Musen Co, Ltd. (Tokyo Japan). Yaesu USA, Yaesu UK, are registered Yaesu companies.

Microsoft and Windows are registered trademarks of Microsoft Corporation.

All other products or brands mentioned in the book are registered trademarks or trademarks of their respective holders.

The Radio Today guide to the Yaesu FTDX101

Table of Contents

Approach 1
Conventions 2
The Yaesu FTDX101D and FTDX101MP 3
Setting up the radio 9
Operating the radio 66
Memory groups and channels 102
Touchscreen display functions 105
Display Soft Keys 110
Function menu 115
Radio settings 125
CW Settings 137
Operation settings 142
Display settings 151
Extension settings 153
Front panel controls 155
Front panel connectors 169
Rear panel connectors 172
Troubleshooting 179
Glossary 188
Table of drawings and images 195
Index 196
The Author 199
Quick Reference Guide 200

OTHER BOOKS BY ANDREW BARRON

The Radio Today guide to the Yaesu FT-710

The Radio Today guide to the Yaesu FTDX10

The Radio Today guide to the Yaesu FT-991A

The Radio Today guide to the Icom IC-705

The Radio Today guide to the Icom IC-7300

The Radio Today guide to the Icom IC-7610

The Radio Today guide to the Icom IC-9700

The Radio Today guide to the Xiegu X6100

The Radio Today guide to the Malahit DSP2

Work the world with D-Star

Work the world with System Fusion

Work the world with DMR

Using GPS in Amateur Radio

Testing 123. Measuring amateur radio performance on a budget

Amsats and Hamsats. Amateur radio and other small satellites

Software Defined Radio

Ham Radio Friedrichshafen. The ham radio exhibition in color or B&W.

ACKNOWLEDGEMENTS

Thanks to my wife Carol for her love and support and to my sons James and Alexander for their support and their insight into this modern world. Thanks also to Yaesu who produced the excellent FTDX101 transceiver and finally, many thanks to you, for taking a chance and buying my book.

ACRONYMS

The amateur radio world is chock full of commonly used acronyms and TLAs (three letter abbreviations :-). They can be very confusing and frustrating for newcomers. I have tried to expand out any unfamiliar acronyms and abbreviations the first time that they are used. But I have assumed that anyone buying an elite class radio such as the FTDX101 will be familiar with commonly used terms such as VFO, AGC, SSB, CW, MHz, and kHz. Near the end of the book, I have included a comprehensive glossary, which explains many of the terms used throughout the book. My apologies if I have missed any.

Approach

Rather than duplicate the Yaesu manual which describes each button, function, and control, I have used a more functional approach. This is a "how to do it" book. For example, I describe how to set up the transceiver for SSB operation. Then I follow that up for CW, FM, RTTY, PSK, and external digital mode software such as FT8. Some people complain it is a "rehash" of the manual. I would like to point out that this book is a user guide to the FTDX101. The Yaesu manual is also a user guide to the FTDX101. Expect some similarities! If I left out some functions or controls, I am sure some people would complain about a lack of completeness.

The aim is not to replace the manual but to more fully explain how to configure and operate the radio to take advantage of its many great features. For example, when I cover the front panel controls, I explain not just what the control does, but how and when to use it. Some of the buttons are disabled in some situations. Along the way, I offer a few 'tips' on how I configured my radio. You don't have to follow these suggestions, but they provide some guidance.

The FTDX101 requires some initial configuration, especially if you want to use external digital mode software such as WSJT-X (FT8), Fldigi, MixW, CW Skimmer, or MRP40. There is a range of Color options for the spectrum display. I imagine that you will experiment with some settings more than once before you decide on the optimum settings for your radio.

I cover updating the radio firmware, loading the Windows driver software for the USB cable connection, and connecting the radio to your PC for CAT control. I even cover the FM mode, how to set up tones for repeater access, and store repeater channels in the radio memory slots.

The 'Setting up the radio' chapter is followed by information about 'Operating the radio' in various modes including using Split with two receivers, VC Tune, and the various noise reduction and notch filters. Then I cover some ways of operating the radio including programming and using the voice keyer and the CW, PSK, and RTTY message keyers, and a few tips on using the radio for the FT8 mode.

The Troubleshooting section deals with some oddities that might trip you up.

The chapters about the front panel controls and the front and rear panel connectors supplement the information provided in the Yaesu manuals. Adding more information about what each control does and how to use them.

Finally, there are chapters describing each of the Soft Key functions and the FUNC menu settings.

The Glossary explains the meaning of the many acronyms and abbreviations used throughout the book and the Index is a great way of getting directly to the topic you are looking for.

Conventions

The following conventions are used throughout the book.

Front panel knobs and buttons are indicated with a highlight.   TUNE

Touchscreen controls are indicated in uppercase without a highlight. RF-POWER

'Touch' means to briefly and gently touch the item on the touchscreen. There is no need to press hard on the screen. Most touchscreen controls are indicated with a beep as well as changing the function or opening a window or menu.

Touch and hold means to keep touching the screen icon until a secondary function or popup window is displayed. This is usually accompanied by a double beep.

'Press' means to press a physical button or knob. The action is usually indicated with a beep as well as changing the function or opening a window or menu.

'Press and hold' or 'hold down' means to hold a physical button or control down for one second until the function changes. It usually opens a control option or sub-menu window on the touchscreen. A press and hold action is usually accompanied by a double beep.

A 'Soft Key' is an icon on the touchscreen that represents a button or control.

Words in <brackets> indicate a step in a sequence of commands. The sequence usually begins with pressing a <button> or a <knob> followed by touching a series of <Soft Keys> or <icons> on the touchscreen.

'Mode' usually means one of the transceiver modulation modes. SSB (USB or LSB), CW, RTTY, PSK, AM, FM, or Data. But it can also mean the 'CENTER', 'CURSOR' or 'FIX' scope display mode, or an external digital mode performed by software running on your PC, such as FT8 or PSK31.

MODE means to press the mode button on the front panel.

FUNC means to press the function button on the front panel.

FTDX101 means that the information applies to both the FTDX101D and the FTDX101MP.

The Yaesu FTdx101D and FTdx101MP

Congratulations on buying or being about to purchase the fabulous Yaesu FTDX101D or FTDX101MP transceiver. It is always exciting unboxing and learning how to use a new transceiver. The radio has superb technical specifications, and it currently holds the top position on the highly regarded Sherwood Engineering transceiver performance table. This book covers both variants of the transceiver. The two radios are essentially the same, with only a few minor differences which will be pointed out along the way. Where I refer to the FTDX101 the information applies to both models.

The FTDX101 has very modern styling, building on the bold Colors and features introduced with the release of the FTDX5000. The front panel is dominated by a bright full-Color 7" touchscreen display and a very large VFO tuning knob with an outer multi-function tuning ring. The controls will be familiar to users of earlier radios such as the FTDX5000 or the FT1000 series. There are "proper" buttons and knobs for the most often used functions. Other functions such as direct entry of a frequency or adjusting the transmitter output power have been moved onto touchscreen controls.

Under the skin, the radio employs two identical 'Hybrid SDR' (software defined radio) receivers. Each receiver has a superheterodyne stage, down converting the incoming radio spectrum to a 9 MHz I.F. A narrowband SDR 'receives' the I.F. bandwidth and converts the signals down to a 24 kHz second I.F. for the DSP (digital signal processing) stage. The radio is not a direct sampling software defined radio, unlike recent releases from other manufacturers. Yaesu has opted for the hybrid technology to ensure top of the range IMD (intermodulation distortion) performance. The radio does include two 'direct sampling SDR' receivers but they are only used for the 'Scope display' accompanying each receiver.

I am sure that you will be excited about the spectrum and waterfall display if this radio is your first SDR. Yaesu doesn't call it a band scope or a panadapter, simply referring to it as the 'Scope' or 'Display.'

The FTDX101 is a truly exceptional radio for contesting or working DX stations. Each spectrum scope can display a band of frequencies up to 1 MHz wide. More than enough to show contest stations across any of the lower bands, or the pileup waiting to work a DXpedition. The two identical receivers are great for monitoring two bands at once or listening to the stations in the pileup while also monitoring a DX station. You can work out which way the DX station is working through the calling stations and hopefully place your transmitter so that you are next in line.

I love the SSB voice message keyer. It makes calling "CQ Contest" or just saying my callsign repeatedly, so much easier.

You can operate CW, PSK, or RTTY without relying on a connection to digital mode software running on a PC. There are built-in decoders for all three modes. But transmitting is limited to sending five pre-set text messages (macros). Although you can use an external USB keyboard to pre-load the text messages, you cannot use it to send text directly from the keyboard. The CW keyer can send a contest number that increments each time the text macro is sent. Sadly, this was not extended to the PSK or RTTY modes. The inability to send free text severely limits the usefulness of the internal PSK and RTTY decoders and I expect that most digital mode operators will continue to use external digital mode software. However, you could work a DXpedition or operate in a CW contest using the built-in software. Especially if you were only using the CW decoder to confirm callsigns and contest exchange numbers.

I like the ability to display two meters so that you can read forward RF power and SWR (or ALC) at the same time. The 'Filter Function Display' is great. It shows the audio bandwidth including the action of the manual notch filter, contour control, or the tuning of the VC TUNE (variable capacitor tune) preselector. The noise reduction system and noise blanker are excellent.

The band selection and receiver controls such as volume, squelch/RF gain, IF shift and width, and the various noise filters, blankers, notch, and contour filters are all duplicated for the second receiver. The sub-receiver controls are marked with blue text rather than white and are located above the main receiver controls.

The FTDX101 does not have a traditional 'menu' structure. Instead, it has a function screen accessed through the FUNC button. The screen displays a bewildering array of 29 options including several 'Soft Keys' that access deeper menu layers. In most cases, the MULTI knob is used to make selections, or you can just touch the relevant icon on the screen.

I cover some "weird" operating behaviors in the troubleshooting chapter. If you experience something strange, have a look to see if it is covered in that chapter. I am sure that some of these oddities will be addressed in future firmware releases.

The FTDX101 is a 'contest grade' transceiver with a fantastic receiver and virtually every option that you could want. It combines cutting-edge SDR technology with all the controls and features that experienced amateur radio operators expect from a Yaesu transceiver. Setting up the radio is relatively easy. But there are a lot of settings. I will cover most of them as we go through the 'step by step' instructions for setting up the radio and discussions about all the functions and features. I included an alphabetically sorted 'quick reference guide' at the back of the book to make it easier to find your way through the menus and make the changes that you want. There is a lot to learn. So, let's get started.

TECHNICAL FEATURES

The FTDX101 is a hybrid SDR. Both receivers use a superheterodyne stage to convert the incoming signals down to a 9 MHz I.F. (Main RX 9.005 MHz, Sub RX 8.900 MHz). This analog stage includes the front-end attenuator, followed by the VC-tune (variable capacitor) preselector, 15 automatically selected bandpass filters, and the RF preamplifier. A wideband signal taken before the bandpass filters is used to feed the two 'direct sampling SDR' receivers which are used solely for the spectrum scope and waterfall displays. The front-end design is the same as used for the main receiver, but not the sub-receiver, of the FTDX5000 radio. On the FTDX5000 the front-end stage is followed by a second oscillator and mixer, but in the FTDX101 an SDR style ADC (analog to digital converter) and FPGA (field programmable gate array) converts the I.F. passband into a digital signal, which is passed on to a 32-bit Texas Instruments DSP (digital signal processing) stage for filtering and demodulation. Finally, a DAC (digital to analog converter) recovers the audio output signal for the audio line output, speaker, and headphones.

The outstanding 123 dB RMDR (reciprocal mixing dynamic range) and 110 dB 3rd order IMD (intermodulation distortion) dynamic range figures are the result of using an exceptional local oscillator and double balanced mixer for the first stage of the receiver, and the high-performance crystal roofing filters. The local oscillator is a 0.1 ppm high stability TCXO (temperature-controlled crystal oscillator) plus a 'high resolution direct digital synthesizer' operating at 400 MHz. The oscillator output is divided down to the required local oscillator frequency, to produce a local oscillator with a 'best in class' phase noise better than -150 dBc/Hz at a 2 kHz offset.

The receiver's selectivity and 3rd order IMD dynamic range is greatly improved by the sharp narrowband roofing filters working at the 9 MHz I.F. The FTDX101D has 600 Hz, 3 kHz, and 12 kHz roofing filters fitted as standard. 300 Hz and 1.2 kHz filters can be added, but only as a factory-installed (not user installed) option. The FTDX101MP has the same three filters installed, plus a 300 Hz filter on the main receiver only. Again, the 1.2 kHz and second 300 Hz filter can be added as factory-installed options. There are no DSP filter options although you can use the IF width and shift controls.

At the start of the receiver chain, there is a 6 dB attenuator followed by a 12 dB attenuator. These are enabled using the ATT Soft Key on the display. They can be combined in series to provide up to 18 dB of front-end attenuation. They are simple voltage dividers consisting of two resistors each. There is another attenuator associated with the IPO (Intercept Point Optimization) function. This is followed by the bandpass filters and then the two optional preamplifiers. AMP1 has 10 dB of gain compared to the IPO level and AMP2 has approximately 20 dB of gain compared to the IPO level. IPO and the preamplifiers are switched in or out with the IPO Soft Key. The 'normal' setting is AMP1.

There are fifteen bandpass filters in the front-end of each receiver. Rather more than in most transceivers. Ten of these are dedicated to amateur bands and the remaining five are for the general coverage receiver.

The VC-Tune tracking preselector is excellent. I have not used one before and I am very impressed. The technology is an enhancement on the High Q RF micro tune system developed for the FTDX9000. Unlike that system there are no noisy relays, switching coils. In the FTDX101 the preselector uses a stepper motor to tune a variable capacitor. The filter attenuates out-of-band signals by up to 70 dB. The FTDX101D has a VC-Tune preselector on the main receiver. You can add a VC-Tune module to the SUB receiver as a factory-installed option. Preferably when you order the radio. The FTDX101MP has VC-Tune preselectors for both receivers.

The transmitter is also an SDR design with the final frequency being generated directly by a 16-bit DAC with no up-conversion mixer. The 400 MHz 'high resolution direct digital synthesizer' used in the receiver also contributes to the outstanding phase noise characteristic of the transmitter. A clock distributor divides and distributes the clock signals to the transmit blocks including the FPGA and DAC. Transmitter phase noise is an impressive -150 dBc/Hz at a 2 kHz offset.

If you haven't used an SDR transceiver before, you will be impressed by how clean the receiver sounds and you will quickly get used to the advantages of the spectrum and waterfall display. The layout is different from previous Yaesu radios but anyone transitioning from an FT1000MP or other Yaesu radios will be comfortable with the controls.

In some ways, the touchscreen seems like an afterthought. I find it frustrating that it is not as configurable as the screens on competitor's radios. The FIX spectrum display is very badly implemented and there should be a dedicated control for the scope level. Or it should auto-level. Yaesu has realized this. In an early firmware update, they added LEVEL to the list of controls that can be dedicated to the large tuning ring around the VFO knob. Surprisingly, this is a good place for this setting, and I leave it allocated that way most of the time. Scope LEVEL will appear on the MULTI knob, but only until you use the control to adjust something else. You cannot change the speed or the dynamic range of the spectrum display and there is no averaging function. There is no beep or on-screen indication when you tune outside of a ham band. The only change is a faint click as the bandpass filter switches from the ham band to general coverage at the end of a 500 kHz band segment. Another mild annoyance is that mouse control does not allow you to activate the FUNC menu. You have to press the button and then you can use the mouse to make selections.

In fact, the only part of the touch screen that does not react to either a mouse click or touching it with your finger is the MULTI indicator in the bottom right corner. This is the obvious place for a Soft Key that enables the FUNC menu.

The 3DSS (three dimensions signal stream) is a spectrum display that includes signal history, similar to the information on a waterfall display. It is represented as a 3D image tapering back to give an illusion of depth. Yaesu is very excited about this way of representing received signals, but I prefer the traditional spectrum and waterfall display. The 3DSS display looks very flash when the radio is on the display stand at a ham convention or trade show, but I don't believe that it adds much value.

Yaesu is justly proud of the 123 dB RMDR (reciprocal mixing dynamic range) measurement at 2 kHz offset, achieved by the receivers. This test result is better than almost all other amateur radio transceivers. A recent independent lab measured the RMDR at 125 dB on the 20m band which is even better than the figure claimed by Yaesu. The receiver two tone 3rd order IMD test results are excellent, placing the radio in the number one spot on the Sherwood Engineering list (Oct 2020). The transmitter is very good as well, with exceptional transmit 3rd order IMD results and low transmitted harmonic levels.

The IPO (intercept point optimization) mode is designed to improve the receiver's dynamic range and intermodulation performance by bypassing the receiver preamplifiers. This does come at the expense of a 10 dB reduction in the receiver sensitivity. It may improve the signal to noise ratio, especially on the noisy low bands. In the event of a very strong signal causing intermodulation problems, you are recommended to try using IPO first and then add the front-end attenuator if the offending signal is still too strong. The 'normal' operating condition is to select AMP1. Turning on AMP2 adds a second 10 dB amplifier to provide a total of 20 dB gain. You might require the extra gain on the 10m and 6m bands.

I find the rather unique 'Contour' control very useful. It functions as a kind of audio tone control since there are no tone controls or audio equalizers provided for the receivers. It creates a shallow manually tuned notch filter and it can be used in that way to combat weak interfering signals. Or, if you want to use it as a tone control, you can adjust the notch frequency to cut or boost either the high or low audio frequencies. The contour notch frequency is indicated by a blue dip or bump on the filter function display. There are menu settings to adjust the width and depth of the notch or boost it provides. The effect of the filter is easy to hear, and it can be seen on the filter function display, or the AF-FFT display if the audio scope displays have been enabled using the MULTI Soft Key.

Some online commentators say that they don't like touchscreens. This is NOT a radio for them. It is impossible to use the radio without touching the screen. All of the menu structure and even basic functions like changing the output power can only be changed via the touchscreen display.

Features that I like about the radio include,

- The fantastic receiver performance.
- The built-in antenna tuner.
- The five voice messages – brilliant.
- The VFO knob. It is smooth and nicely weighted, with adjustable drag.
- A great squelch system – so I don't have to listen to the band noise.
- The APF (audio peak filter) for CW – it's great!
- The ZIN button to pull CW signals to the correct tone.
- Being able to display RF power output and SWR (or ALC) at the same time.
- The audio 'filter function display.' It shows the audio bandwidth including the action of the notch filter, contour control, or the VC-Tune (variable capacitor). I like it a lot.
- The CW decoder and contest message keyer.

There are a few things that I hope will be upgraded. They include,

- There is no averaging control for the spectrum scope and no dedicated control for adjusting the spectrum level.
- At a normal squelch setting, the Main receiver squelch knob gets in the way of the MULTI control.
- The radio is not able to record signals off the air.
- Not being able to use an internal or external keyboard to send PSK, CW, or RTTY direct from the keyboard. I would settle for some way to quickly insert a station's callsign. That would at least make the function usable. By the way, the same problem exists with other manufacturers' radios which include PSK and RTTY message keyers. I believe that only the Kenwood TS-890 and TS-990 support sending from an external keyboard.
- You cannot configure the radio to change antennas automatically when you change bands. Although the radio does remember the last used antenna, so I guess that is OK.
- You cannot use the mouse to turn on the FUNC menu. It is surprisingly annoying if you are using a mouse to control the radio.
- The FIX spectrum display mode lacks the functionality that I expect. You should be able to select any start and stop frequency with a span of up to 1 MHz. Other transceivers allow four fixed spectrum display ranges per band.

Setting up the radio

The Yaesu manual does a good job of identifying all the controls and menu items, but it lacks detail in some areas. In this chapter, I cover the processes that you need to follow to get ready for operating the radio. Each section lists the process that you need to follow when you set the audio levels, tone settings for FM, and transmitter bandwidth. It is important to set the Mic Gain AMC (automatic microphone gain), and compression controls correctly for SSB operation. I explain the setup for the FM, CW, RTTY, and PSK modes including the built-in options and using external digital mode programs. There are instructions on configuring the five voice messages for SSB and the pre-defined keying messages for CW, RTTY, and PSK.

I tell you how to download and install the USB driver software to allow the radio to communicate with your computer and discuss setting up the virtual COM ports and the Audio CODEC, linear amplifier connections, and the increasingly popular FT8 mode. Finally, I cover formatting and using the SD card, how to do firmware updates, how to set the clock, and the ANT3 antenna switching matrix.

The radio is easy to configure once you know which settings to use. Where I have changed my radio from the Yaesu default settings, I have explained why I made the change and the effect that it has on the radio. The big advantage of using the instructions in this section rather than using the Yaesu manual is that I have included all the necessary steps, and the optional ones, in one place. I tell you what the controls and settings do and how they should be adjusted.

DISPLAY YOUR CALLSIGN WHEN THE RADIO STARTS

It is nice to personalize the radio by having it display your name or callsign as the radio 'boots up.' You can set up to 12 characters to be displayed upon start-up.

Select <FUNC> <Display Setting> <My Call> and use the onscreen keyboard to change the message. There was just enough room for me to put 'ZL3DW FT101D.'

Press ENT when you have finished, to save the changes, and then BACK, BACK, or FUNC to exit.

SETTING THE TIME AND DATE

The radio has a clock, but it is never displayed on the screen. It is only used for time stamping the filename of saved files. Select <FUNC> <EXTENSION SETTING> <DATE&TIME> and set the DAY, MONTH, YEAR, HOUR, and MINUTE.

Tip: If you want files date stamped with UTC date and time, enter UTC date and time details. But I believe it is easier to use local time.

GETTING READY FOR SSB OPERATION

Three settings should be adjusted before you start transmitting on SSB. You need to set the Microphone Gain control, the AMC (automatic microphone gain) control, and the Speech Processor (PROC) level. AMC is an automatic leveling control that prevents you from overmodulating the transceiver. You may not intend to use the speech processor (compressor), but you might as well 'set it and forget it,' so that it is correctly adjusted if you do decide to use it.

The AMC and Mic Gain settings are very broad, and it is difficult to get a satisfactory result. You might prefer to just set the three parameters to the settings in the 'SSB Audio Settings' table below and forget about the adjustment procedure. But if you want to "give it a go," carry on with the steps below.

TIP: These settings also apply to the AM and FM modes, so there is no additional setup required for AM or FM. Although you can set the microphone level on those modes using the<FUNC> <RADIO SETTING> <mode> <MIC GAIN> menu if you want to.

If you want to listen to the transmitted audio to check the quality, or just hear "how it sounds," you can turn on the audio monitor by pressing the MONI button in the row of buttons to the right of the display screen. The monitor level is adjusted using <FUNC> <MONI LEVEL> and rotating the MULTI knob.

TIP: The Yaesu operating manual says that AMC should be adjusted so that the COMP meter never deflects past 10 dB. This seems to be impossible and is causing all sorts of confusion on the forums. My advice is to ignore the statement in the manual and make an adjustment that swings over most of the meter scale rather than sitting sit mostly in the region between 15 dB and full scale. A setting of 75 to 80 seems appropriate.

Typical setup

It is always preferable to make transmitter adjustments with the radio connected to a suitable high power (100W+) 50 ohm load. A high-power (100W+) 50 ohm 50 dB or 60 dB attenuator will work just as well.

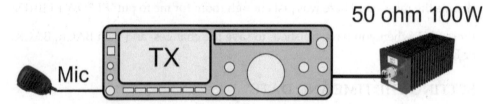

If you do not have a suitable load, you can set up the SSB levels while transmitting into your normal antenna. Preferably do it during the daytime when the HF bands are closed and announce your callsign regularly. You can set the SSB levels on any HF band provided you have a suitable RF load or antenna.

Setup

1. If the radio is not already set for SSB operation, press the SSB button to select USB assuming the frequency you are using is above 10 MHz. Select LSB if you are using one of the three low bands.

2. Select a two-meter display by ensuring that the MONO Soft Key is blue and the EXPAND Soft Key is white. You can display two meters in other display modes, but this way works well. If you end up with one meter and one DSP filter function display, touch the filter spectrum to enable the second meter.

3. Touch either of the meters and select COMP on the left meter and ALC on the right meter.

4. Check that the radio has the adjustment of the Speech Processor level enabled. Press <FUNC> <OPERATION SETTING)> <TX AUDIO> <PROC LEVEL> <COMP>. If it is set to AMC, you cannot set the Speech Processor compression level. Although AMC is the default setting, I believe that this setting should always be set to COMP. Press FUNC to exit the 'Operating Setting' screen.

   Note that changing this setting to AMC does not turn off the Speech Processor (COMP) function. It only prevents you from adjusting the compression level. See 'Confusion about AMC and Speech Compressor operation' below.

5. If the orange LED under the word PROC next to the MIC/SPEED control, is lit, press the MIC/SPEED knob to turn off the Speech Processor.

6. The Mic Gain and the AMC Out level controls interact with each other. If the initial Mic Gain is too high, or too low, you will not be able to set the AMC Out level correctly. If the AMC Out level is too high, or too low, you will not be able to adjust the Mic Gain appropriately. Use the inner MIC/SPEED knob to set the Mic Gain to an initial value of 35.

7. Use the outer control labeled PROC/PITCH to initially set the AMC OUT level to 80. You can make adjustments from there. The control works backwards. At 100 there is no AMC levelling protection. At levels under 50, there is so much audio leveling that the transmitter power is reduced. Some people prefer an AMC setting less than 55. In fact, they are quite aggressive about it. See 'Setting the AMC OUT level below 55,' on page 14.

8. It is a personal preference, but I believe that it is easier to adjust the AMC setting if the AMC RELEASE TIME is set to FAST. Otherwise, the COMP meter takes ages to settle after each voice peak. You can always slow the action back down to MED or SLOW after you have finished adjusting the transmitter. Use <FUNC> <OPERATION SETTING)> <AMC RELEASE TIME> <FAST>.

9. Check that the radio will be transmitting into a suitable high power 50 ohm load or a resonant antenna. You can turn down the RF power on your FTDX101MP to suit a 100 Watt load.

Method

10. Key the transmitter with the microphone PTT switch and talk in your normal 'on-air' voice. In all the excitement is tempting to talk louder or shout. But you want to modulate at the same level you will be using when you are using the radio. I usually say my callsign and "testing 12345 54321" a few times. You could use a typical CQ format, but someone may come back to your call. Click the inner MIC/SPEED knob a couple of clicks anti-clockwise and check that the display shows a popup box that says MIC GAIN. Watch the ALC meter on the right side and adjust the inner MIC/SPEED knob so that the voice peaks are reaching the top area of the white ALC range and not peaking into the blue range. Don't worry if the average peaks are not reaching the top of the ALC scale. Peaking to about the 10 mark on the I_D scale below the ALC range should be fine. A whistle should peak the meter to just under 15.

11. The microphone level may be lower than you were expecting. My level ended up at 30. Much lower than that and the ALC meter won't peak to the top of the white ALC range. Much higher and the COMP meter will be peaking up near 20 dB, which is too high.

    TIP: The level of the microphone gain is not particularly critical. If it is set too high the AMC will work harder to make sure that the transceiver never gets overloaded. I did not see any clipping distortion at any setting.

12. With the left meter set to COMP and the AMC OUT set between 75 and 80, the COMP meter should spend most of the time between 5 and 10 dB, with occasional peaks up to 20 dB or higher. Even with the AMC RELEASE TIME set to FAST, the COMP recovers quite slowly after peaks. So, I just left it set to FAST.

➢ **Setting the speech processor compression level**

It is worthwhile setting the Speech Processor compression level even if you do not plan on using the compressor. "Set and forget."

Setup

This adjustment carries on from the previous adjustments. The setup is identical.

Method

13. Press the MIC/SPEED knob to turn the Speech Processor on. This is indicated with an orange LED just to the left of the control knob.

14. Turn the outer PROC/PITCH knob a click or two to confirm that the popup now says PROC LEVEL. If it does not, you need to change the PROC LEVEL menu setting to COMP. (Step 4 above).

15. Check that the left side meter is set to COMP.

16. While transmitting into a suitable RF load or resonant antenna, key the transmitter with the microphone PTT switch and talk in your normal 'on-air' voice.

17. While observing the compression indicated on the COMP meter, use the PROC/PITCH knob to adjust the PROC LEVEL. Yaesu recommends that the compression level be kept under 10 dB. I have my PROC LEVEL adjusted for normal speech to be peaking between 5 dB and 10 dB. With very loud peaks to a little over 15 dB. I ended up with a PROC LEVEL setting of 15.

SSB AUDIO SETTINGS		
Function	ZL3DW Setting	My Setting
MIC GAIN	30	
AMC OUT	80	
PROC LEVEL	15	

➢ Confusion about AMC and Speech Compressor operation

The description of the AMC and Speech Compressor settings in the Yaesu manual is rather cryptic and this has led to a lot of confusion.

The 'PROC LEVEL' setting <FUNC> <OPERATION SETTING)> <TX AUDIO> <PROC LEVEL> (COMP or AMC), is also mentioned in the 'FTDX101D Important Changes.pdf' document.

1. If the LED under the words MIC/SPEED and PROC is lit, the Speech Processor (compressor) is turned on. This is irrespective of the <FUNC> <OPERATION SETTING)> <TX AUDIO> <PROC LEVEL> (COMP or AMC) setting in the FUNC menu.

2. Since a Firmware upgrade, (possibly earlier than V10-14), the AMC (automatic microphone gain) is always active and cannot be turned off. Although it can be turned down using the AMC OUT control.

3. If the 'PROC LEVEL' setting is set to AMC. The PROC/PITCH (outer) knob will adjust the AMC OUT level, even if the Speech Processor (compressor) is turned on. This setting leaves you with no way to adjust the speech processor COMP level. So, it is better to leave it set to the COMP option.

4. If the 'PROC LEVEL' setting is set to COMP and the Speech Processor is turned on, the PROC/PITCH (outer) knob will adjust the Speech Processor compression 'PROC LEVEL.'

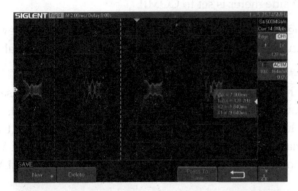

Figure 2: **SSB no compression** Yaesu FTDX101D 20m band. Test by author

Figure 1: **SSB with compression** Yaesu FTDX101D 20m band. Test by author

> **Setting the AMC OUT level below 55**

Adrian (VK4), Tim P (K4), and Tim M all advocate using an AMC OUT level of less than 55. They acknowledge that this causes neither of the Yaesu setup conditions to be met but they insist that there is a Yaesu instruction to this effect. So far, nobody on the Yaesu group forum has been able to produce a published Yaesu document that states that the AMC OUT level should *"NEVER exceed 55"* as claimed, so it must be a well-guarded secret. Setting the AMC level to 55 is fine. I have no problem with anyone doing that. It will cause the ALC meter to read very low, and the COMP meter, with the speech processor turned off, i.e. reading the AMC OUT level, to read very high. On my transmitter, I am unable to output more than 50 watts at that AMC setting. But your transmitter and voice may be different.

Part 1 on page 50 of the Yaesu manual states, *"Key TX and adjust the [MIC/SPEED] knob to set the input level of the Microphone Amplifier to the position where the ALC meter does not exceed the ALC zone on audio peaks."* Setting the AMC OUT level to 55 results in a low ALC meter reading that does not move much on voice peaks.

I ran a check of the transmitter audio dynamic range. It is reduced by 2 dB at an AMC level of 55 and 3 dB with an AMC level of 50. This will probably not be noticed by a receiving station; you will just sound a little less lively.

Page 50, part 2 states, *"press the [MIC/SPEED] knob so that the indicator is off"*, *"Adjust the AMC to a point where the COMP Meter does not exceed "10 dB" on the audio peaks."* Setting the AMC OUT level to 55 results in a very high COMP meter reading that peaks well over 20 dB.

I am quite surprised that a Yaesu employee, would say on the Yaesu group forum, that if you set the AMC level so that the leveller is off, or at a level higher than 55, the radio is likely to create "splatter," causing interference to other band users, possibly breaching FCC and local spectrum rules, and forever blackening the Yaesu brand. He says, *"I think factory default on AMC is either 50 or 55? I can't remember. Many claim to be running AMC levels "higher" than 55. I wouldn't recommend it personally, nor does Yaesu recommend it. I did for awhile (as I said), and got good audio...with.....lots of products / whiskers / splatter. Of course, many told me the "usual"......"oh, your audio sound fine". But the more critical/technical people looked at there bandscopes and said "you've got issues". All that went away when I properly set the AMC (below 55)."*

I don't care! You can set the AMC OUT level to anything you like. Please don't email and tell me about it.

AMC TESTS

Since I have been rather rudely flamed over this issue and told, *"You just don't understand how it works,"* and *"You are spreading misinformation via your book."* I have performed a series of tests. Using real test instruments rather than gossip.

- **Power output and ALC test**

I input a 1 kHz tone to the transmitter with; mode USB, REAR input, COMP off, and AMC OUT initially set to 100. I set the audio input level for an ALC reading in the gap between the white and blue zones on the ALC meter because that is the maximum mic level that Yaesu recommends. *"Adjust the [MIC/SPEED] knob to set the input level of the Microphone Amplifier to the position where the ALC Meter does not exceed the ALC zone on the audio peaks,"* i.e. this is the **maximum** level you would expect a voice peak to reach.

Then I recorded the Po power output and ALC meter reading as I reduced the AMC OUT level in 5 unit steps. For the ALC reading, I recorded the numbers on the I_D scale since the ALC meter does not have a scale. Initially, the AMC OUT control has no effect, and then as the limiter begins to work the ALC level drops progressively. When the ALC reaches near zero it can no longer keep the transmitter power at 100% and the RF power starts to drop.

Of course, at that point, you can increase the Mic gain to compensate, but the transmitted audio will still have a small reduction in dynamic range. For the test, I continued to turn down the AMC level. The RF output continues to drop.

Eventually, the AMC OUT number gets so low (heavily levelled) that even very high Mic Gain settings cannot get the radio to transmit at full power. Note that the test was done with a single tone. The average level of a speech signal is lower, so the drop off in ALC and transmitter power may occur earlier, (at higher AMC levels).

The RF power reading (dashed line) uses the scale on the left side of the chart and the ALC reading (grey solid line) uses the scale on the right of the chart.

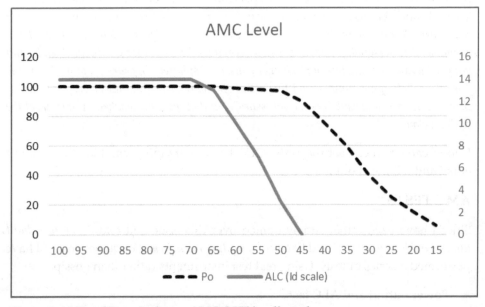

Figure 3: **ALC and Po metering as AMC OUT is adjusted**

The good news is that I did not see any evidence of envelope distortion at any AMC setting. I doubt that Yaesu would allow the transceiver to be set at a level that causes RF splatter! Surely that would be irresponsible.

From the chart, it seems that any setting above 70 is pretty much the same unless you have a very high input level. My setting of 80 is conservative and the guys that set their radios to 100 are fine. Looking at the chart I would say that the correct setting is where the ALC starts to drop. i.e. 65 to 70. However, because speech is different to a single tone, if you want the ALC meter to reach the top of the white area on speech, you will probably have to set the AMC at around 75 to 80. At lower settings you will have reduced ALC readings, (not in itself a bad thing), and you may have to increase the MIC gain a lot to get 100W (200W on an MP) RF power output.

I cannot get 100 watts output at any microphone gain setting with the AMC set to 55 or below. But if you like QRP that's fine by me. Your voice or transmitter may be different.

I checked the COMP meter reading (with the speech processor turned off) as well, but it just follows the gain of the AMC OUT level so there are no surprises there.

The COMP reading is heavily dependent on the type of modulation. With a single tone, the reading stays at zero until the AMC reaches 70 then increases to full scale as the AMC OUT control continues to be reduced. With a two tone audio signal, the increase is fairly linear, starting when the AMC is 100 and not levelling off until the AMC is at 20. And with speech input, the COMP meter starts reading when the AMC is at 100 and flattens out at a high level when the AMC levels are below 55.

➢ **2 tone IMD and RF spectrum checks**

I was really rather concerned about the claim by Tim P, Tim M, and Adrian, that the transceiver would produce splatter if the AMC OUT level was set higher than 55. I could not readily accept that Yaesu would allow that to happen. But here was a Yaesu employee clearly stating that if you follow the setup instructions in the Yaesu manual you will be transmitting a distorted signal, with *"lots of products / whiskers / splatter."*

I carried out 2 tone IMD tests with a 500 Hz offset and again with a 1 kHz offset with the AMC set to 100, 80, and 55 (to make Adrian happy). I also checked the RF spectrum on speech. I am pleased, but not surprised, to report that the 2 tone IMD figures are substantially the same at all three AMC settings. The 3rd order product is slightly better at 100 than it is at 55, but only 1.29 dB, so not anything of concern. The 5th order product is actually worse than the 3rd order product, which is unusual, but I have seen it with other SDR transmitters. The 5th order product is slightly better at 55 than it is at 100, (probably because the output power is reduced by the same margin), but the delta is only 2.25 dB so not anything of concern. The radio easily meets both the ARRL and ITU-R SM.326-7 recommendation for IMD performance, at any AMC setting. YAY!

Many forum members have said that they use AMC settings of 80 to 100 and have received excellent reports. My test was carried out at 14.188 MHz on USB, with the default transmitter bandwidth setting. I checked the transmitted spectrum at AMC settings of 100, 80, and 55 (to make Adrian happy). I am not surprised, to report that I saw no evidence of RF splatter at any AMC setting. The RF skirt is about 3 dB lower with the AMC set to 55, but the RF output power is lower by a similar amount, so the results are similar in terms of dBc.

The spectrum scope displays 5 dB per division. Turn the keyer off and BK-IN on, then calibrate the spectrum display to the PEP level by setting a CW signal to the top of the spectrum display using the LEVEL control. See page 187 for the process.

SETTING THE TRANSMITTED BANDWIDTH

You can set the bandwidth of the transmitted signal for the SSB, AM, or PSK/DATA modes.

Press <FUNC> <RADIO SETTING)> <MODE SSB> <TX BPF SEL>

Press <FUNC> <RADIO SETTING)> <MODE AM> <TX BPF SEL>

Press <FUNC> <RADIO SETTING> <MODE PSK/DATA> <TX BPF SEL>

Since firmware V01-20, the default for SSB is 100-2900 Hz (2.8 kHz BW), but I set my radio to 200-2800 Hz (2.6 kHz BW) because I work contests and DX. The old default of 300-2700 (2.4 kHz BW) is a bit too narrow for most operators.

For AM, I left my radio at the default 50-3050 Hz (3 kHz BW) option because it seems reasonable, and I never use AM anyway.

I changed the Data mode bandwidth to 100-2900 Hz (2.8 kHz BW) because it almost matches the bandwidth of the 200-3000 Hz receiver passband displayed on the digital mode software. The 50-3050 Hz option may fit better, but I would not be comfortable transmitting a digital mode signal below 100 Hz.

PARAMETRIC EQUALIZER

You can tailor the audio frequency response of your transmitted signal using the three-band parametric equalizer. There are settings for when you have the compressor turned off and for when you have the compressor turned on.

Before setting up the microphone equalizer, you have to turn it on. The setting is a toggle, and the function menu will disappear the instant you touch the MIC EQ Soft Key.

- Press <FUNC> and look at the <MIC EQ> Soft Key. If it says OFF, touch <MIC EQ> to turn it ON.
- Make sure that the speech compressor is turned off. The orange LED labeled PROC beside the MIC/SPEED knob should be off. If it is on, press the knob to turn it off.
- Set the transmitter bandwidth (as per the previous section).
- The parametric equalizer settings are at <FUNC> <OPERATION SETTING> <TX AUDIO>.

It is always preferable to make transmitter adjustments with the radio connected to a suitable high power 50 ohm load.

A high-power 50 dB or 60 dB, 50 ohm attenuator will work as well. The setup is the same as for the SSB adjustments above.

It is a good idea to listen to your transmitter audio quality using headphones. You can use the speaker in the radio, but headphones are much better for this task.

Turn on the transmit monitor using the MONI button. Press and hold the MONI button for a double beep, then turn the MULTI knob to adjust the volume. This does not change the function associated with the MULTI control. Or you can use <FUNC> <MONI LEVEL> which will associate monitor level with the MULTI knob. The audio for the transmit monitor is taken from the transmitter I.F. so it is a good representation of the signal that is being transmitted. Alternatively, you can listen to another receiver.

The equalizer settings are a personal preference. However, I would avoid using very narrow bandwidth settings because they will make the audio passband lumpy. You are aiming for a fairly linear slope across the audio passband. There is no point in placing the low-frequency filter below the low edge of the transmitter bandwidth or the high frequency above the high end of the transmitter bandwidth. The equalizer will have no effect on frequencies that are not within the band of frequencies being transmitted.

FREQ sets the center frequency of the low, mid, or high filter

LEVEL sets the gain or loss at the center frequency in dB (+10 to -20 dB)

BWTH sets the bandwidth of the filter (0 to 10)

➢ **Speech processor (compressor) turned off**

PARAMETRIC EQUALISER SETTINGS – Compressor Off			
Control	Function	ZL3DW Setting	My Setting
PRMTRC EQ1 FREQ	LOW FREQ	300 Hz	
PRMTRC EQ1 LEVEL	GAIN	-6 dB	
PRMTRC EQ1 BWTH	Q (Bandwidth)	Q 5	
PRMTRC EQ2 FREQ	MID FREQ	900 Hz	
PRMTRC EQ2 LEVEL	GAIN	0 dB	
PRMTRC EQ2 BWTH	Q (Bandwidth)	Q 5	
PRMTRC EQ3 FREQ	HIGH FREQ	2400 Hz	
PRMTRC EQ3 LEVEL	GAIN	+8 dB	
PRMTRC EQ3 BWTH	Q (Bandwidth)	Q 5	

➢ **Speech processor (compressor) turned on**

The speech processor (compressor) only works in SSB mode, so the 'P PRMTRC' settings only work when you are using SSB, and you have the speech processor on.

TIP: Page 52 item 3 of the Yaesu manual says, "To adjust the Parametric Microphone Equalizer with the AMC or Speech Processor engaged, press the [MIC/SPEED] knob to activated (sic) AMC or Speech Processor." This is misleading. It should say, "if you want to use the Speech Processor press the knob so that the LED is lit." There is a similarly misleading statement in point 3 of the right- side column.

Make sure that the compressor (speech processor) is turned on. The orange LED labeled PROC beside the MIC/SPEED knob should be on. If it is off, press the MIC/SPEED knob to turn it on.

PARAMETRIC EQUALIZER SETTINGS – Compressor On			
Control	Function	ZL3DW Setting	My Setting
P PRMTRC EQ1 FREQ	LOW FREQ	300 Hz	
P PRMTRC EQ1 LEVEL	GAIN	-6 dB	
P PRMTRC EQ1 BWTH	Q (Bandwidth)	Q 5	
P PRMTRC EQ2 FREQ	MID FREQ	900 Hz	
P PRMTRC EQ2 LEVEL	GAIN	0 dB	
P PRMTRC EQ2 BWTH	Q (Bandwidth)	Q 5	
P PRMTRC EQ3 FREQ	HIGH FREQ	2400 Hz	
P PRMTRC EQ3 LEVEL	GAIN	+8 dB	
P PRMTRC EQ3 BWTH	Q (Bandwidth)	Q 5	

SETTING UP THE VOICE MEMORY KEYER

The voice memory keyer is a very useful feature of the radio. It can be used for sending CQ calls or any information that you transmit a lot. It is especially useful for contest operation. Especially if the contest has a fixed 'contest exchange' such as the signal report and your CQ zone number. You can record the contest exchange, "*thanks you are 59 32*" as well as your "*CQ Contest.*" Having one message with just your callsign in phonetics, "*Zulu lima three delta whisky*" is useful for when you are trying to break the pile-up to work a rare DX station or a DXpedition.

There are five voice memory slots. Each recording can be up to 20 seconds long. It is longer than you think. Mine are all less than 10 seconds long.

Recording or sending the voice messages is done using the menu command <FUNC> <REC/PLAY>. This brings up the 'Message Memory' popup. Unfortunately, the box disappears if you touch the screen or press any button and it overlays the center of the spectrum and waterfall display. I can see why investing in the FH-2 external keypad is a good idea.

FH-2 keypad: You can use the FH-2 keypad, or similar, to record or send voice messages. You cannot use a USB keyboard.

Recording: Touch the MEM Soft Key. A red REC indicator will flash at the top-right of the Message Memory popup window.

Touch one of the message keys, 1 to 5. When you are ready to record, click and release the microphone PTT switch. The REC indicator will stop flashing. Record your message in your 'normal' on-air voice. Touch MEM to complete the recording.

Do the same thing to record the other 4 macros. Touch BACK or, the screen outside of the popup window, or any button to exit.

RX Level: Sets the volume of the message as it plays back. You will hear the recording even if you have MONI turned off. If it annoys you, turn down the RX Level to zero. The default level was very loud. I have set mine at 10%.

TX Level: Sets the level to be transmitted. You should set it so that it is approximately the same as when you say the same message using the microphone. I looked at the COMP and ALC meters while transmitting and set the TX LEVEL for similar meter readings as when I use the microphone. I am sure that 50% would be fine, but I ended up setting TX Level to 60% on my radio.

Playback: Once the messages have been recorded, you can touch 1 – 5 to play them back. You can listen to your recordings by keying any of the messages with BK-IN turned off. You must have BK-IN turned on to transmit the messages.

Bug: Firmware releases before V01-20 (April 2021) had a bug affecting the audio quality when listening to the message with BK-IN turned off. See Troubleshooting.

SETTING UP THE CW MESSAGE KEYER

There are two ways you can record messages for the CW keyer. You can type in the macro as TEXT from the on-screen or external USB keyboard, or you can record a MESSAGE sent with a CW Paddle.

Before you record or type the message you have to set the five message memory slots to TEXT or MESSAGE. You can have some slots set to TEXT and others set to MESSAGE. Select TEXT or MESSAGE for each of the five memory slots using <FUNC> <CW SETTING> <KEYER> <CW MEMORY>

TIP: It could be handy to have one slot set to MESSAGE so you can use the paddle to quickly save another station's callsign, during or immediately before beginning a QSO.

If you have the FH-2 keypad or similar, you can press MEM to initiate a recording from the keypad. This is useful if you are using the MESSAGE mode and recording from the paddle.

> **Saving a CW keyer memory (TEXT method)**

To save text to a CW keyer memory slot. Make sure the message slot you want to use is set to TEXT. You can have some slots set to TEXT and others set to MESSAGE.

Press <FUNC> <CW SETTING> <KEYER> <CW MEMORY (1-5)> <TEXT>.

- Set the radio to CW mode.
- Turn BK-IN off.

- Press \<FUNC\> \<REC/PLAY\> \<MEM\> and select message 1-5.

A keyboard will appear on the display. Type in your message. You are allowed up to 50 characters. The last character must be the } symbol. It is used to tell the transceiver to finish transmitting and return to receiving. The arrow keys move you across the message. The 'X in an arrow' symbol is the backspace. Touch ENT to save the message, or BACK to exit without saving.

TIP: To listen to the message without transmitting, set BK-IN to off and then touch the message number. BK-IN must be turned on for the message to be transmitted.

To insert a **contest number**, enter a # character. For example, 'R 5NN # K}'. The contest number will be sent as a three-digit number with leading zeros. The contest number will extend to four digits once 1000 is reached.

You can **decrement** the contest number by touching the DEC Soft Key. To set the number back to 1 use \<FUNC\> \<CW SETTING\> \<KEYER\> \<CONTEST NUMBER\>.

The number format is set using the \<FUNC\> \<CW SETTING\> \<NUMBER STYLE\> command. See page 140 for the different options.

Morse code prosigns

For KN use	(	Over to a specified station
For AS use	&	Stand By
For AR use	+	End of message
For BT use	=	New section or paragraph
For CT use	%	New message
For SN use	!	Message understood

As far as I can tell, you cannot enter prosigns such as AA, BK, BN, CL, HH, or SK into a CW text memory. It should be possible to enter them from the paddle if you are using the MESSAGE method.

> **Saving a CW keyer memory (MESSAGE method)**

First, make sure the message slot you want to use is set to MESSAGE. Press \<FUNC\> \<CW SETTING\> \<KEYER\> \<CW MEMORY (1-5)\> \<MESSAGE\>.

To save a message sent by a CW paddle to a CW keyer memory slot. Set the radio to CW mode. Turn BK-IN off. Turn MONI on, unless you are using an external keyer with its own sidetone. Press \<FUNC\> \<REC/PLAY\> \<MEM\>. Slots set to MESSAGE show a series of vertical bars instead of text. Touch to select one of them.

A red REC icon will flash. Start sending your message within 10 seconds and touch MEM to stop the recording.

TIP: To listen to the message without transmitting, set BK-IN to off and then touch the message number. BK-IN must be turned on for the message to be transmitted.

TIP: You could allocate one memory slot for editing during or immediately before a QSO to put in the other station's callsign.

➢ **Sending a message**

Once recorded, you can send the message by pressing 1-5 on the FH-2 keypad, or by opening the message memory popup, <FUNC> <REC/PLAY> and touching 1-5. Like all CW operation, you must have BK-IN turned on. The text string is displayed on the bottom line of the decoder screen if it is turned on. Each character turns yellow as the text message is sent.

➢ **Stopping a message from being sent**

To stop a message while it is being sent, start sending from the paddle, touch the message memory key on the touchscreen, or press the message key on the FH-2 keyer.

GETTING READY FOR CW OPERATION

The main controls for CW operation are the keying speed and pitch. They can be adjusted at any time, to suit your needs.

➢ **Morse keys**

There are provisions for different types of key. The front and rear panel 'Key' jacks can be used with a Paddle, Bug, or Straight Key. Note that you must turn on BK-IN for the keyer to automatically place the transceiver into transmit mode. The menu settings are discussed in the CW SETTING menu on page 137.

➢ **CW Keyer**

You activate the internal keyer by setting the radio to CW mode and pressing the MIC/SPEED knob so that the orange 'Keyer' LED is on. Once this has been done the keyer will be active anytime you select the CW mode.

See CW SETTING – KEYER sub-menu on page 139 for the full list of keyer settings.

➢ **Keying speed and pitch**

When the radio is in CW mode and the keyer is turned on, the MIC/SPEED knob sets the speed of the electronic keyer including CW sent from the CW message keyer. As you adjust the control, the CW speed in words per minute (wpm) is displayed on a popup in the middle of the display. The CW speed is adjustable from 4 wpm to 60 wpm.

The outer PROC/PITCH control changes the pitch of a received CW signal without changing the receiver frequency. You can set the control so that CW sounds right to you. A popup displays the pitch frequency. It is adjustable from 300 Hz to 1050 Hz. Most people use 700 to 1000 Hz. I prefer 700 Hz.

> **Pitch offset on SSB**

CW signals are received at an offset equal to the 'pitch' setting of the PROC/PITCH control. The behavior when you switch from CW to SSB varies depending on whether you have selected the default 'pitch offset' or the 'direct frequency' option. Note that this setting only affects the VFO in SSB mode. In CW mode the VFO always shows the frequency of the received CW signal.

<FUNC> <CW SETTING> <MODE CW> <CW FREQ DISPLAY> <PITCH OFFSET>

<FUNC> <CW SETTING> <MODE CW> <CW FREQ DISPLAY> <DIRECT FREQ>

The default PITCH OFFSET setting offsets the VFO when you switch from CW to SSB, so you hear the CW signal at the same tone. If you select the DIRECT FREQUENCY setting, the VFO frequency will not be offset. If you change to SSB, the CW tone will be at the zero-beat frequency and will not be heard.

I believe that the default PITCH OFFSET setting is preferable. It means that you can tune in a CW signal while on SSB and then switch to CW without losing the signal.

> **Break-in setting (BK-IN) button**

Your CW Morse Code signal will not be transmitted unless BK-IN has been turned on. You can leave it on all the time. The setting is remembered when you change modes. With BK-IN off you will hear the CW signal but it will not be transmitted unless the MOX button, the microphone PTT, or an external PTT signal perhaps from a footswitch, is activated.

There is no front panel control for selecting full or semi break-in. I guess most people use one or the other most of the time.

Select SEMI or FULL Break-in using <FUNC> <CW SETTING> <MODE CW> <CW BK-IN TYPE>.

- Full break-in mode will key the transmitter while the CW is being sent and will return to receive as soon as the key is released. This allows for reception of a signal between CW characters. There is some relay clicking but it is not too intrusive.
- Semi break-in mode will key the transmitter while the CW is being sent and will return to receive after a delay when the key is released. The break-in delay is set using the next menu setting. <FUNC> <CW SETTING> <MODE CW> <CW BK-IN DELAY>.

➢ Roofing filter

Use the R.FIL Soft Key (or the R.FIL menu command) to set the roofing filter. It will normally be set to 600 Hz for CW. The FTDX101MP also has a 300 Hz roofing filter, on the main receiver only.

➢ The DSP filter function display

The DSP filter function display indicates the currently selected roofing filter and other things such as the manual notch, contour control, APF filter position, I.F. width, and I.F. shift.

In CW mode a bar graph under the spectrum display indicates the tuning of the CW signal. This provides yet another method of tuning in the signal correctly. When the highlighted three dots are right in the middle under the red dot, the signal is tuned perfectly.

You can turn the bar graph on or off with <FUNC> <CW SETTING> <MODE CW> <CW INDICATOR>

➢ Sidetone

Turn on the transmit monitor MONI button to hear the CW sidetone. The sidetone level is adjusted by selecting <FUNC> <MONI LEVEL> and using the MULTI knob to set the transmit monitor level. Or you can press and hold MONI to make a quick adjustment. Changing the monitor level on CW will not affect the monitor level when you are in SSB mode.

➢ CW Keyer messages (keyer macros)

There are five CW message memory slots which can be used for DX or Contest operation or just to save you sending the same message over and over. They are great for sending CQ on a quiet band. The CW keyer can send an auto-incrementing contest number. See 'setting up the CW message keyer' on page 23 in the previous section.

➢ APF filter

Don't forget to use the fabulous APF (audio peak filter). It is an audio DSP filter centered on the CW pitch frequency. Because the APF is so narrow it is very effective at lifting weak CW signals out of the noise. There are three APF bandwidths available. The wider bandwidth setting will allow for the VFO being slightly off tune, but in a contest situation, you could hear more than one CW signal. The narrow setting requires that the signal be tuned exactly, or you may "lose the signal."

<FUNC> <OPERATION SETTING> <RX DSP> <APF WIDTH>

If there is another CW signal very close to the wanted signal, you can offset the tuning of the APF using the CONT/APF control. Turning the control in CW mode automatically turns on the APF filter.

The (600 Hz) receiver filter passband and any offset of the APF filter are shown on the DSP filter function display. APF is indicated with a vertical red line.

➢ **CW keying in SSB mode**

Some people like to use a CW sign-off when operating SSB. The CW AUTO MODE menu setting <FUNC> <CW SETTING> <MODE CW> <CW AUTO MODE> lets you enable or disable the Morse key or paddle when you are on SSB. When set to ON the key is 'live' but CW will only be transmitted if the BK-IN switch is enabled (LED lit). When set to OFF, the key is disabled on SSB. If you select 50M the key is active on the 6m band but not on the HF bands. Unless you really want the ability to send CW while in SSB mode, I suggest leaving this function turned off.

➢ **QSK delay time**

<FUNC> <CW SETTING> <MODE CW> <QSK DELAY TIME> sets the time before the transceiver will start sending the CW signal. This allows your linear amplifier to switch before any power is transmitted.

The default is 15ms which should be fine for any modern solid-state amplifier. However, you can select a longer delay if your amplifier uses relay switching. If you are operating at more than 45 wpm the radio automatically selects 15 ms regardless of the menu setting.

RTTY AND PSK MESSAGE KEYERS

The RTTY and PSK message keyers are identical, and nearly the same as the CW keyer. There is no way to send a contest number and you use an 'End' ↵ character to finish the text string instead of the } symbol. A contest number would have been handy. I have no idea why the RTTY and PSK message keyers are different in this way. There are five message memories. Each can hold 50 characters.

➢ **Saving or editing a message**

To save text to a PSK or RTTY keyer memory slot.

- Set the radio to PSK or RTTY-L mode.
- Press <FUNC> <REC/PLAY> <MEM> and select message 1-5.

A keyboard will appear on the display. Type in your message. You are allowed up to 50 characters. The last character must be the ↵ symbol entered by touching 'End.' The 'End' ↵ symbol is used to tell the transceiver to finish transmitting and return to receiving. The arrow keys move you across the message. The 'X in an arrow' symbol is the backspace. Touch ENT to save the message, or BACK to exit without saving.

TIP: You could allocate one memory slot for editing during or immediately before a QSO to put in the other station's callsign.

➢ **Sending a message**

Once recorded, you can send the message by pressing 1-5 on the FH-2 keypad, or by opening the message memory popup, <FUNC> <REC/PLAY> and touching 1-5. The text string is displayed on the bottom line of the decoder screen. Each character turns yellow as the text message is sent.

Unlike the voice keyer and the CW keyer, you do not have to have BK-IN turned on to transmit a saved message.

➢ **Stopping a message from being sent**

To stop the message while it is being sent, touch the message memory key on the touchscreen, or press the message key on the FH-2 keyer.

GETTING READY FOR RTTY OPERATION

The radio supports three kinds of RTTY operation. Firstly, there is the onboard RTTY decoder. Which can be used with the RTTY message memories. The second method is to use external PC software such as MixW, MMTTY, DM780, or Fldigi with AFSK (audio frequency shift keying). AFSK uses two audio frequencies to create the frequency-shift keying in the SSB DATA-U mode. The third method is to use external PC software with FSK, which uses digital signals to key the transceiver to predefined mark and space offsets. FSK uses the radio's RTTY-L or RTTY-U mode, so you can use the RTTY message memories.

➢ **Keyer send messages (keyer macros)**

There are five RTTY message menus which can be used for DX or Contest operation or even just to save you sending the same message over and over. They are great for sending CQ on a quiet band. The setup is on page 26 in the previous section.

The keyer messages only work if you are in the RTTY mode, so they will work for the internal RTTY mode and when using an external RTTY program with FSK keying, but not if you are using the DATA-U mode and an external AFSK RTTY program.

➢ **Onboard RTTY operation**

Using the onboard RTTY mode is simple. You have an onboard RTTY decoder and a set of five pre-defined messages. Unfortunately, the operation of the transmit messages is not integrated with the decoder screen.

1. Select the RTTY-L mode by holding down the MODE button until the selection window pops up. The polarity setting in the FUNC menu should already be set to NOR (normal).

2. Turn on the decode screen. <FUNC> <DECODE>. This must be done before activating the RTTY keyer. The decoder only has two controls. DEC LVL sets the decoding threshold. Touch the Soft Key and adjust the DECODE LVL with the MULTI knob to reduce the amount of rubbish text being displayed. Note that the MULTI control reverts to whatever it was on after 4 seconds.

   The other Soft Key closes the window. The five blank Soft Keys should be used for sending the five RTTY messages but they are not.

3. Below the received text area there is one line reserved for displaying the outgoing text. The characters turn yellow as they are sent.

4. If you have the FH-2 external keypad or a copy, use that to send the message macros. Otherwise, turn on the RTTY keyer popup. <FUNC> <REC/PLAY>. Touch 1 to 5 to send the appropriate message.

   The popup covers almost all of the decoded text. YAY! Touching any part of the screen outside of the popup or pressing some of the buttons closes the popup and you have to go back through the FUNC menu to get it up again.

TIP: If the decoder is printing gibberish the other station might be transmitting on the wrong sideband. Switch to RTTY-U to decode the signal. Amateur band RTTY should be on RTTY-L. Commercial Shortwave band RTTY is likely to be received on RTTY-U.

➢ AFSK RTTY from an external PC program

To use an external PC-based digital modes program for transmitting you use the DATA-U mode. Not the RTTY mode. Press and hold the MODE button until the selection popup appears. Then select DATA-U.

The easiest way to send RTTY from your favorite external digital mode program is to use AFSK rather than FSK keying. With AFSK, the RTTY signal is sent to the transceiver as audio tones rather than a digital signal. I have included setup instructions for several digital modes programs in the 'Setting up PC and digital mode software' section.

First of all, you have to have a connection between the transceiver and the PC. The easiest and most modern method is to use a single USB cable. It will carry the CAT control commands, COM port lines for PTT and keying, and the audio. See the section below on Setting up a USB connection on page 34. It only has to be done once. You can also use the old-fashioned separated communications and audio cables method, but I have not covered the setup for that.

➢ AFSK RTTY audio levels

Once the digital mode software can communicate with the radio, it is time to set the audio levels on transmit and receive. You can set the audio level for the transmitter in the FTDX101, or you can set the levels using the soundcard mixer controls in your PC or possibly with level controls in the digital mode software.

<FUNC> <RADIO SETTING> <MODE PSK/DATA> <RPORT GAIN>

I prefer to leave it at the default setting of 50 and use the Windows Sound settings or controls inside the digital mode software to adjust the levels.

There is no control for setting the level of the receiver audio output to the PC via the USB cable. You will have to use the audio controls on the PC software or in Windows.

If you have not already set up the USB cable and audio settings, see 'Setting up a USB connection' on page 34.

➢ **FSK RTTY**

I have included instructions for setting up MMTTY for FSK RTTY in the section titled, 'Setting up PC and digital mode software.' MMTTY is often used in conjunction with the N1MM logger. FSK also works with Fldigi. I cannot get FSK to work with DM780, the digital modes program that is used by Ham Radio Deluxe. However, CW and the AFSK RTTY mode work OK.

➢ **RTTY Mark and Shift frequencies**

You can change the RTTY MARK tone frequency and the RTTY shift. The default Mark tone is at 2125 Hz and the Shift when using RTTY-L is +170 Hz.

You can change the settings, but I don't see any reason to do it.

<FUNC> <RADIO SETTING> <MODE RTTY> <MARK FREQUENCY>

<FUNC> <RADIO SETTING> <MODE RTTY> <SHIFT FREQUENCY>

➢ **RTTY polarity settings**

You can change the RTTY polarity settings for the receiver and for the transmitted RTTY signal. The default for amateur radio is to leave them set to NOR.

<FUNC> <RADIO SETTING> <MODE RTTY> <POLARITY TX>

<FUNC> <RADIO SETTING> <MODE RTTY> <POLARITY RX>

With a NOR setting the VFO indicates the Mark frequency and the data signal causes transitions to the Space frequency.

With a REV setting the VFO indicates the Space frequency and the data signal causes transitions to the Mark frequency.

➢ **NOR, REV, RTTY-L, and RTTY-U**

It is easy to confuse the normal and reverse polarity settings set in the RADIO SETTING menu with the RTTY-L and RTTY-U settings selected with the MODE button.

For amateur radio, we use the RTTY-L and NOR settings. Commercial Shortwave band RTTY is likely to be received on RTTY-U.

If a station is transmitting RTTY on the wrong sideband you can switch to RTTY-U or REV to receive the signal correctly.

RTTY Settings		
Function	Operation	Tones
NOR	The VFO indicates the Mark frequency	FSK data shifts the frequency from the Mark tone to the Space tone
REV	The VFO indicates the Space frequency.	FSK data shifts the frequency from the Space tone to the Mark tone
RTTY-L	Mark 2125 Space 2295 (+170)	Transmits Mark low, Space high
RTTY-U	Mark 2295 Space 2125 (-170)	Transmits Mark high, Space low

GETTING READY FOR PSK OPERATION

The radio supports two kinds of PSK operation. Firstly, there is the onboard PSK decoder. Which can be used with the PSK message memories. The second mode is used with external PC software such as MixW, MMTTY, DM780, or Fldigi.

The menu settings for PSK operation are included in the chapter about the FUNC menu commands, so I won't repeat them here.

<FUNC> <RADIO SETTING> <MODE PSK/DATA> (on page 131)

<FUNC> <RADIO SETTING> <ENCDEC PSK> (on page 134)

Select the PSK mode by press and holding the MODE button. You will see that there is no upper or lower sideband selection because BPSK (binary phase-shift keying) is sent using 180-degree phase changes rather than tones like RTTY and the sideband does not matter. However, QPSK (quadrature phase-shift keying) must be received on the same sideband it was transmitted on, so there are normal and reverse polarity settings in the menu structure. <FUNC> <RADIO SETTING> <ENCDEC PSK> (on page 134)

BPSK-31 and QPSK-31 are supported by the internal PSK encoder and decoder. BPSK-63 and QPSK-63 are not. Not a big problem since PSK-63 is rarely used anyway.

If you want to change the PSK tone for the internal PSK mode, select <FUNC> <RADIO SETTING> <MODE PSK/DATA> <PSK TONE>. You can choose from 1000 Hz (default), 1500 Hz, and 2000 Hz. This setting only affects the audio frequency you will hear when you tune into a PSK signal. It is like the audio offset applied to a CW signal.

➢ Onboard PSK operation

Using the onboard PSK mode is simple. You have an onboard PSK decoder and a set of five pre-defined messages. Unfortunately, the operation of the transmit messages is not integrated with the decoder screen.

5. Select the PSK mode by holding down the MODE button until the selection window pops up.

6. Turn on the decode screen. <FUNC> <DECODE>. This must be done before activating the PSK keyer. The decoder only has two controls. DEC LVL sets the decoding threshold. Touch the Soft Key and adjust the DECODE LVL with the MULTI knob to reduce the amount of rubbish text being displayed. Note that the MULTI control reverts to whatever it was on after 4 seconds. The other Soft Key closes the window. There are five blank Soft Keys that should be used for sending the five RTTY messages but are not.

7. If you have the FH-2 external keypad or a copy, then use that to send the message macros. Otherwise, turn on the PSK keyer popup. <FUNC> <REC/PLAY>. Touch 1 to 5 to send the appropriate message.

The popup covers a large part of the decoded text. YAY! Touching any part of the screen outside of the popup or pressing some of the buttons closes the popup and you have to go back through the FUNC menu to get it up again.

➢ PSK message keyer (keyer macros)

There are five PSK message menus which can be used for DX or Contest operation or even just to save you sending the same message over and over. I have already covered the setup back in a previous section (page 26).

The keyer messages only work if the radio is in the PSK mode. The inability to send messages from a connected keyboard, or even enter the other station's callsign without saving it to one of the five memory slots is rather limiting. So, I expect most people will use an external digital modes program for general PSK operation.

➢ **PSK from an external PC program**

To use an external PC-based digital modes program for transmitting you use the DATA-U mode or DATA-L, not the PSK mode. Press and hold the MODE button until the selection popup appears. Then select DATA-U.

First of all, you have to have a connection between the transceiver and the PC. The easiest and most modern method is to use a single USB cable. It will carry the CAT control commands, COM port lines for PTT and keying, and the audio. See the section below on Setting up a USB connection. It only has to be done once. You can also use the old-fashioned separated communications and audio cables method, but I have not covered the setup for that. See page 34 for the USB cable setup.

After the USB settings, I have included typical COM port settings for a selection of popular digital mode programs. I can't include them all, but the settings I describe should be representative of what you need.

➢ **PSK audio levels**

Once the digital mode software can communicate with the radio, it is time to set the audio levels on transmit and receive. If you have already set the RPORT audio level for the PSK/DATA mode, you are done. Go and work some PSK stations. If not, carry on reading.

RPORT GAIN sets the audio level into the transmitter from the USB port or RTTY/Data jack. You could turn it down so that the transmitter cannot be over-modulated.

But I decided to leave it at 50 and adjust the modulation level using the PC (Windows) sound card levels and/or the level control on the external software or connected device. You can jump ahead to page 131 for the full table of PSK settings.

<FUNC> <RADIO SETTING> <MODE PSK/DATA> <RPORT GAIN>

There is no control for setting the level of the receiver audio output to the PC via the USB cable. You will have to use the audio controls on the PC software or in Windows.

➢ **BPSK and QPSK compared**

You can choose from BPSK (binary phase-shift keying) which is standard PSK-31, or QPSK (quadrature phase-shift keying). I assume that it is QPSK-31 since there is no baud rate control offered. QPSK has the advantage of having built-in error correction, but it is considerably less sensitive than BPSK, and much less popular on the bands. <FUNC> <RADIO SETTING> <ENCDEC PSK> <PSK MODE>.

Binary PSK transmits data information by changing the phase of the carrier signal. A digital one is represented by a 180-degree phase change and a digital zero is represented by no phase change. The 31 baud rate of PSK-31 is chosen to provide a narrow bandwidth transmission at a typical typing speed. These days there is little PSK activity since most of the world has moved to FT8. The faster PSK-63 mode was less popular because you need a higher received signal to noise ratio to get error-free reception. In other words, it is less sensitive. It also uses twice the bandwidth.

QPSK is much faster than BPSK because each of the four-phase states carries two data bits. The higher symbol rate allows time to insert some additional 'information bits' to enable the use of forward error correction, so you tend to get nearly perfect decoding or none at all.

QPSK requires more signal level for good decoding. It has never really caught on as a general chat mode, and most people that do use it prearrange their contacts. Typically, QSPK is transmitted on frequencies just above the PSK segment of the band.

BPSK-31 is a narrow bandwidth mode like CW, so it works well at low signal strengths. Not as well as FT8 of course. It also works well on radio paths with fading or trans-polar paths with auroral flutter and multi-path effects. Ideal for 'long path' contacts between New Zealand and northern Europe.

The internal PSK decoder has AFC (automatic frequency control) to help you tune the PSK signal accurately. It will pull the receiver slightly to get the signal tuned in. <FUNC> <RADIO SETTING> <ENCDEC PSK> <DECODE AFC RANGE> sets the bandwidth for the AFC action. A BPSK-31 signal has a bandwidth of about 31 Hz. So, I would experiment with the 30 Hz or 15 Hz settings.

GETTING READY FOR FT8 AND OTHER DIGITAL MODES

Before you use the radio with WSJT-X for FT8 or other digital mode software you need to set up a connection between the radio and your PC. This is easiest using a USB cable, because transmitter and receiver audio and rig control, (including PTT switching, CW signaling, and FSK RTTY) can all be carried over a single cable. See the next section 'Setting up a USB connection' about how to load the driver software onto your computer and establish the COM port connections. Then you need to set up the software on the PC. There are hundreds of software options. I have included basic setup instructions for some of the most popular packages in the 'Setting up PC digital mode software' on page 41.

➢ **PRESET**

The April 2021 V01-20 firmware release added a 'Preset' mode to the MODE button. The idea is to provide an easy way of changing the transceiver settings for FT8 or other digital modes, then changing it back when you have finished.

NOTE: you still have to set DATA-U. The preset does not include mode switching. Touch and hold the MODE button and set the mode (DATA-U for FT8). Then touch and hold the MODE button again to turn on the PRESET. *Could this be more clunky?*

When the PRESET Soft Key is blue the preset settings are being applied. When the PRESET Soft Key is grey the preset settings are not being applied.

Warning! On my radio attempting to adjust the 14 presets by touch and holding the PRESET Soft Key crashes the radio and I have to power down to reset it. There is a work-around. Change <S.MENU> <PEAK> to LVL1 before entering the PRESET setup menu. See Troubleshooting, "FT8 PRESET locks up transceiver," on page 182.

The preset setup uses a standard menu screen, which makes me wonder why Yaesu has not added the PRESET setup to the RADIO SETTING tab on the FUNC menu, instead of making it the only touch and hold Soft Key on the MODE selection screen.

Touch and hold the PRESET Soft Key to change the 14 preset settings.

Function	PRESET Settings (**default**)	ZL3DW	My settings
CAT RTS	**OFF** or ON	OFF	
CAT RATE	4800 / 9600 / 19200 / **34800**	19200	
CAT TIME OUT	**10ms**/100ms/1000ms/3000ms	10 ms	
LCUT FREQ	OFF / **100 Hz** to 1 kHz	100 Hz	
HCUT FREQ	OFF 700 to 4000 Hz (**3200 Hz**)	3200 Hz	
TX BPF SELECT	50-3030 Hz / 100-2900 Hz/ 200-2800 Hz/ 300-2700 Hz/ 400-2800 Hz	50-3050 Hz	
REAR SELECT	DATA/**USB**	USB	
RPORT GAIN	0 to 100 (**10**)	10	
RPORT SELECT	DAKY / **RTS** / DTR	RTS	
AGC FAST DELAY	20 – 4000 ms (**160 ms**)	160 ms	
AGC MID DELAY	20 – 4000 ms (**500 ms**)	500 ms	
AGC SLOW DELAY	20 – 4000 ms (**1500 ms**)	1500 ms	
LCUT SLOPE	6 dB/octave or **18 dB/octave**	18 dB	
HCUT SLOPE	6 dB/octave or **18 dB/octave**	18 dB	

SETTING UP A USB CONNECTION

To use PC software to control the radio you have to connect a USB 2.0 Type B to Type-A cable between the USB port on the rear of the radio and a USB 2.0 or 3.0 port on the PC. Feel free to use a spare USB 2.0 port on your computer as there is no advantage in wasting a USB 3.0 port.

Using the USB cable for the first time is **not** "plug-n-play." You must download and install the Yaesu virtual COM port driver software.

Setting up the radio | 35

> **Driver software**

The Yaesu driver software can be downloaded from the Yaesu website https://www.yaesu.com/. Select the FTDX101D or FTDX101MP in the 'Products' tab or click the appropriate radio on the main screen. Click on the 'Files' sub-tab. In the third section, 'Amateur Radio \ Software,' click and download the file labeled 'FTDX101D/MP USB Driver (Virtual COM Port Driver).' There is an installation guide for it, but you won't need it because you have me.

The link will download a ZIP file. In Windows, click 'Extract All' and save the files to a suitable place on your computer. I created a 'Yaesu FTDX101D' folder under 'My Documents.' Then you can delete the ZIP file. Once the files have been unzipped, the ZIP file is only taking up hard drive space.

Navigate to the location where you saved the driver files. For Windows 10, double click the file called CP210xVCPInstaller_x64.exe to start the installation. It will install the 64-bit version of the driver. If you are running a 32-bit X86 operating system run the 32-bit installer. Follow the onscreen instructions to complete the installation.

After the driver has been installed. Connect the PC to the transceiver using the USB cable and turn the radio on. You should get the usual device connected 'bong' from Windows.

In Windows 10 select 'Settings' and then 'Devices.'

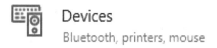
Devices
Bluetooth, printers, mouse

Under the 'Other Devices' heading, you should see a new device with the rather catchy name of, 'CP2105 Dual USB to UART Bridge Controller.' If the device is not listed, the PC is not seeing the radio. Try unplugging the USB cable from the PC end and then plugging it back in again. If it still won't work, reload the Yaesu driver software, with the USB between the radio and the PC connected and the radio turned off.

The driver software creates two COM ports. You will need to know the COM port numbers when you set up digital mode or other PC software. The 'Enhanced' COM port is used for CAT control of the radio. The 'Standard' COM port cannot be used for CAT control, but its RTS and DTR lines are used to key the radio to transmit (PTT) and to send CW or FSK RTTY data.

36 | The Radio Today guide to the FTDX101

Start Windows 'Device Manager' by typing 'Device Manager' into the 'Find a setting box' on the same Windows screen or the Windows search box next to the 'Start' icon. Expand the 'Ports (COM and LPT)' line, to show the installed COM ports.

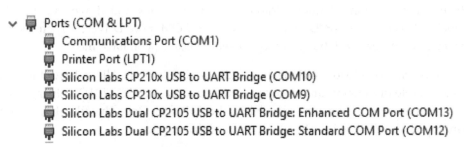

Figure 4: Installed Windows COM ports

The COM ports of interest are the 'Silicon Labs Dual CP2105 USB… **Enhanced** COM Port' and the 'Silicon Labs Dual CP2105 USB… **Standard** COM Port.' Make a note of the COM port numbers and make sure you know which one is the Enhanced port.

➢ **Audio Codec**

In addition to creating the two USB ports, the driver software creates an audio CODEC (coder-decoder). This makes the radio look to the computer like a sound card or an audio device, like a microphone or speakers.

In Windows 10 select 'Settings' and then 'Devices.'

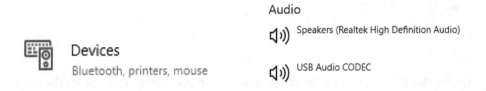

You should see the Yaesu audio codec listed as 'USB Audio CODEC.' If it is not there, the sound won't work. Try unplugging the USB cable from the PC end and then plugging it back in again.

Click on 'Sound Settings' (right side of the window), then.

Click on 'Sound Control Panel' (right side of the next window).

Transmitter audio: Under the **Playback tab** you will see a device labeled as 'USB Audio CODEC Speakers' or something similar. I found the labeling confusing when trying to pick the right device in my digital modes programs, so I changed the name to 'FTdx101D TX.'

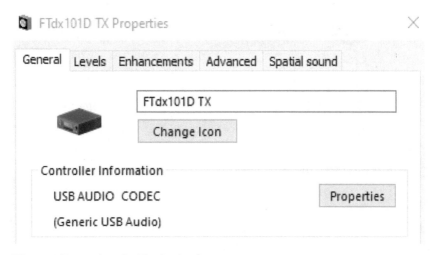

Figure 5: Renaming the Playback tab

The Playback tab controls the audio output from the PC into the transmitter. Double click the icon and change the name on the General tab.

You can pick a different icon as well if you like. Click on the 'Levels' tab and reduce the audio level to about 40. Make sure there are no 'Enhancements' or 'Spatial Sound' effects selected. On the 'Advanced' tab, select 16 bit 48000 Hz (DVD Quality) and check both 'Exclusive mode' check boxes. Press the 'OK' button to return to the 'Sound' window.

Receiver audio: The **Recording tab** controls the audio output from the receiver to the computer. You will see a device labeled as 'USB Audio CODEC Microphone' or something similar. I found the labeling confusing when trying to pick the right device in my digital modes programs, so I changed the name to 'FTdx101D RX.'

Double click the icon and change the name on the General tab. You can select 'Change Icon' to pick a different icon as well if you like.

Click on the 'Levels' tab and reduce the audio level to about 25. On the 'Listen' tab make sure that 'Listen to this device' is not checked. The dropdown should be on 'Default Playback Device.'

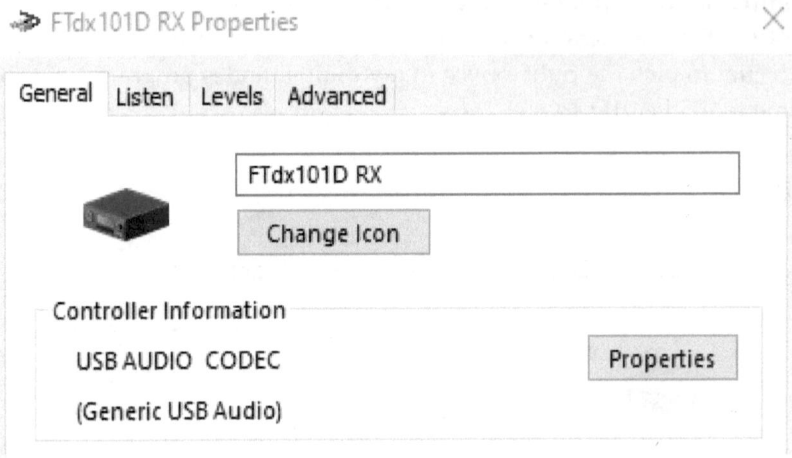

Figure 6: Renaming the Recording tab

The default setting on the Advanced tab should be 2 channel 16 bit 48000 Hz (DVD Quality) or 2 channel 44100 Hz (CD Quality). It must be set to a '2 channel' option.

The main receiver is on the Left audio channel and the sub receiver is on the Right audio channel. Note that the transceiver volume controls do not affect the audio level sent over the USB cable, but the squelch controls are active.

➢ **USB cable settings for the DATA-U and DATA-L modes**

The following settings relate to the USB cable connection when using the DATA modes. They only need to be set once. See page 131 for a detailed explanation.

<FUNC> <RADIO SETTING> <MODE PSK/DATA> <DATA MOD SOURCE> <REAR>

<FUNC> <RADIO SETTING> <MODE PSK/DATA> <REAR SELECT> <USB>

<FUNC> <RADIO SETTING> <MODE PSK/DATA> <RPORT GAIN> <50>

<FUNC> <RADIO SETTING> <MODE PSK/DATA> <RPTT SELECT> <RTS>

➢ **USB cable settings for the PSK mode**

The settings for PSK are the same as the DATA mode settings above. If you have set them, you are already done. See page 131 for a detailed explanation.

➢ **USB cable settings for the RTTY mode**

There are no USB audio settings for FSK RTTY because it is a keyed mode (like CW).

➢ **Radio settings for the SSB mode (external voice)**

The following settings relate to the USB cable connection when using the SSB modes. They only need to be set once. See page 125 for a detailed explanation.

Setting up the radio | **39**

SSB should not be used for digital modes from an external program. Use the DATA-U mode instead.

<FUNC> <RADIO SETTING> <MODE SSB> <SSB MOD SOURCE> <REAR>

This setting would only be used if you were sending speech from the PC. Otherwise, leave it set to MIC.

Note that the microphone will work normally even if this is set to REAR. But the voice keyer will not work. With BK-IN on, the radio will go to transmit, but the message will not modulate the transmitter.

<FUNC> <RADIO SETTING> <MODE SSB> <REAR SELECT> <USB>

<FUNC> <RADIO SETTING> <MODE SSB> <RPORT GAIN> <50>

<FUNC> <RADIO SETTING> <MODE SSB> <RPTT SELECT> <RTS>

➢ **COM Port settings in the FTDX101 - CAT settings**

Other than the COM Port settings discussed above, three menu settings affect CAT operation.

CAT RATE is used to set the interface speed of the USB connection between the radio and the PC software.

The speed should match the speed specified in the software application. It does not need to be very fast. 9600 bps (bits per second or 'bauds') is adequate for most connections. The default rate is 38400 bps.

<FUNC> <OPERATION SETTING> <GENERAL> <CAT RATE> <38400 bps>

CAT TIME OUT TIMER sets the time that the radio will wait for a response to a CAT command. Leave it at the default setting unless you are experiencing CAT control problems.

<FUNC> <OPERATION SETTING> <GENERAL> <CAT TIME OUT TIMER> <10>

CAT RTS is not the PTT control line. That is set elsewhere and relates to the second COM port. In this case, it is used as a 'com interrupt.' When it is set to 'on' the radio monitors the RTS line on the CAT COM port and will respond to changes initiated by the PC software. When it is set to 'off,' the radio does not monitor the RTS line on the enhanced COM port.

<FUNC> <OPERATION SETTING> <GENERAL> <CAT RTS> <ON>

➢ **COM Port settings in PC software**

The COM port settings for the PC are usually set in each digital mode or other PC programs. They can also be set in Device Manager, but that does not seem to be necessary. These are settings I use.

Enhanced Port (used for CAT control)

- Port   COM x (the Enhanced CAT com port)
- Baud Rate   38400
- Data bits   8
- Parity   None
- Stop bits   1
- RTS   Hardware, Handshake, or PTT
- DTR   Hardware, Handshake, or CW

RTS and DTR must not be set to OFF.

Standard Port (used for PTT and CW signaling)

- Port   COM x (the Standard CAT com port)
- Baud Rate   9600
- Data bits   8
- Parity   None
- Stop bits   1
- RTS   PTT (must be the same as the RPTT setting)
- DTR   CW (must be the same as the PC KEYING setting)

The RTS and DTR labels don't matter. You can use either line for the transmit PTT as long as you use the other line for CW. The names relate to old-fashioned RS-232 communications between 'old school' computers. RTS stands for 'ready to send' and DTR stands for 'data terminal ready.' But the lines have not been used for that sort of signaling since the 1970s.

Anyway, I use 'ready to send' for the 'PTT' command because when PTT is active, you are ready to send the digital or CW signal. That leaves 'data terminal ready' for the CW or RTTY data signal.

➢ **USB cable audio settings**

RPORT GAIN sets the audio level into the transmitter from the USB port or RTTY/Data jack. You could turn it down so that the transmitter cannot be over-modulated. But I decided to leave it at 50 and adjust the modulation level using the PC (Windows) sound card levels and/or the level control on the external software or connected device. There is an RPORT GAIN control for SSB (page 125), AM (page 127), FM (page 129), and DATA (page 131). There is no RPORT GAIN control for CW or RTTY because they are 'keyed' modes.

It is not possible to set the receiver output level to the computer. So, you have to use the controls in the PC software.

The receiver volume controls do not affect the audio level being sent over the USB cable to the computer software. But the squelch does. This means that you cannot have the receiver squelched and still send audio to the computer. I find this a bit disappointing. I would prefer that the audio to the computer over the USB cable be independent of the squelch but there does not seem to be any way to change it.

> **External digital mode software - audio levels**

Once the digital mode software can communicate with the radio, it is time to set the audio levels on transmit and receive. If you have already done that for another digital mode or other PC software, it will be fine for FT8 as well.

I have chosen to leave the RPORT audio level at the default 50% setting and will adjust the audio levels on the PC if required. See the images on the next page for typical settings.

Figure 7: My Playback level FTDX101 TX

Figure 8: My Recording level FTDX101 RX

SETTING UP PC DIGITAL MODE SOFTWARE

Having set up a USB connection between the computer and the radio you will need to configure the digital modes or other software. This can be tricky and naturally, it is not covered in the Yaesu manual. In this section, I have included setup information that "works for me" for some popular digital mode and control programs. But of course, it is impossible to cover them all, and I have not covered all the settings, just the basic COM port ones. If you get stuck. Try Googling the application name and FTDX101 and hopefully some help will be on the way. There are also some very good YouTube clips and the Yaesu users groups at https://groups.io/groups. Some of my digital modes software does not have a setting for the FTDX101 and this can be a big problem because the CAT commands are not always the same. I tried FT991 and FT5000 settings with mixed success.

➢ Downloads

I strongly advise downloading the latest version of any software you want to configure to work with the FTDX101. It can save you hours of frustration. If it still doesn't work, at least you know you are using the latest version when you go online to seek help.

➢ N1MM Logger+

I am using N1MM Logger+ V 1.0.8708. It has support for the FTDX101. To set up communications with the radio, start N1MM with the transceiver turned on.

Enhanced Com port: Under 'Config - Configure Ports, Mode Control, Audio, Other - Hardware.'

- Set the Port dropdown list to the Yaesu Enhanced COM port.
- Set the Radio dropdown to FTDX101D.
- Uncheck CW/Other because you are not using this port for PTT.
- Click 'Set.'
- I have speed = 38400 (same as the radio), Parity = N, Bits = 8, Stop = 1, DTR = Handshake, RTS = Handshake. Radio = 1. Nothing else is selected.

Setting up the radio | 43

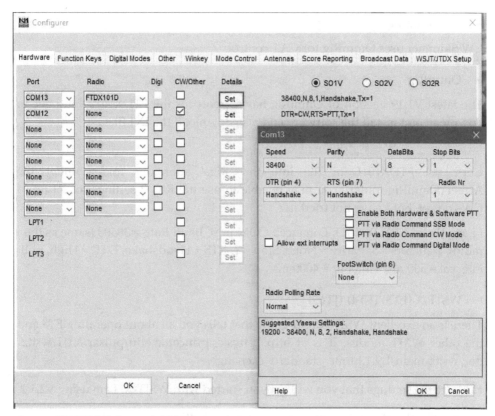

Figure 9: N1MM Enhanced COM port setting

Standard Com port: Under 'Config - Configure Ports, Mode Control, Audio, Other - Hardware.' Set the second Port dropdown list to the Standard COM port.

Set the Radio dropdown to None. Check CW/Other because you are using this port for PTT. Click 'Set.' I have set DTR = CW, RTS = PTT, VFO = 1, PTT Delay = 30 ms. Nothing else is selected.

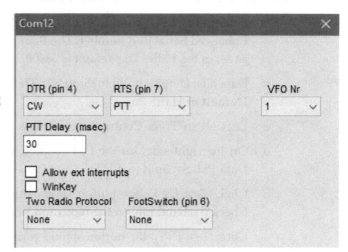

Figure 10: N1MM Standard COM port setting

➢ CW Skimmer

CW Skimmer uses OmniRig for CAT control.

➢ OmniRig

The latest V1.19 version of OmniRig has support for the FTDX101. You have to download and install the software and also download the INI files.

http://www.dxatlas.com/Download.asp

Copy the FTDX101D.ini file into the 'Rigs' subfolder folder under the Afreet\OmniRig folder. You may be able to use the RIG 2 settings for PTT and CW control, but I have not tried it.

I have Radio = FTDX101D, Enhanced COM port, Baud Rate = 38400 (same as the radio), Data Bits = 8, Parity = None, Stop = 1, RTS = Handshake, DTR = High, poll interval = 500 ms, Timeout = 4000 ms.

➢ WSJT-X (FT8) (FT4) (JT65)

There is an excellent WSJT-X user guide that tells you all about operating FT8 and the other WXJT modes. It is at http://physics.princeton.edu/pulsar/K1JT/wsjtx-doc/wsjtx-main-1.9.1.html#_standard_exchange.

Here are the settings that you will get you started with WSJT-X. I am using V2.1.2.

- On the WSJT-X software, go to 'File' 'Settings' 'General' and set your callsign, six-digit Maidenhead grid, and IARU region.
- Go to 'File' 'Settings' 'Radio' and set Rig to Yaesu FTDX101D. (Only V1.9 and later versions of WSJT-X have the FTDX101 option).
    - Under the 'CAT Control' heading, set the Serial Port to the Enhanced Serial port number. The **Baud Rate** should be the same as set in the radio. The default is 38400, but I am using 19200.
    - Data bits: **Default** or Eight, Stop bits: **Default** or One, Handshake: **Default** or Hardware.
    - Leave both 'Force Control Lines' dropdown boxes blank.
    - On the right side, set the PTT method to RTS. Set the Mode to Data/Pkt. Set Split to 'Rig.'
    - Click 'Test CAT,' the button should turn green. If it does not. Check the baud rate on the radio and WSJT-X.
    - Click 'Test PTT,' the button should turn red and the transmitter should go to transmit, but no transmitter power should be created. Don't forget to click the button again to turn off the transmitter.

- Go to 'File' 'Settings' 'Audio'
    - Set the Input to your FTDX101 audio output. Set the other setting to 'Left.' (The Right audio channel carries audio from the sub-receiver). For some reason, Windows reports the Input as 'Line (USB AUDIO CODEC)' even though I relabeled it in Windows to 'FTDX101 RX.' No other software does this.
    - Set the Output to your FTDX101 audio input, FTDX101 TX. Leave the other setting at Mono or Left to match the receiver. It does not matter.
- Click OK to exit.

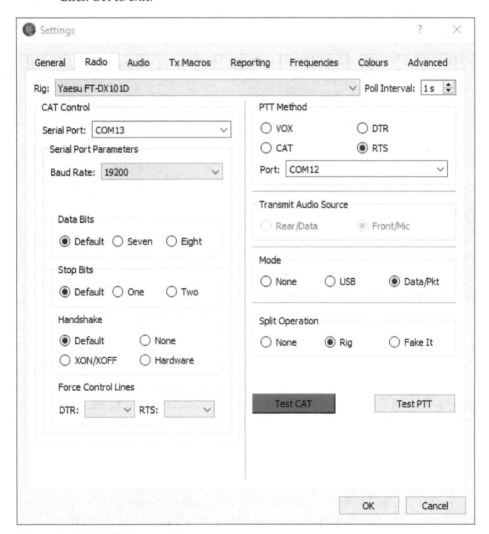

Figure 11: WSJT COM port settings

At this stage, the radio frequency should be indicated on the WSJT display and it should change if you turn the radio's main-VFO knob. You should be able to click the band dropdown beside the frequency display and select a band. The radio should follow your selection.

> **JS8Call**

The setup for 'JS8 call' is very similar to the setup for WSJT. You create a configuration file for the FTDX101 by cloning or renaming the 'default' setting. Select your new Config file and then click 'File' then 'Radio.'

On the 'CAT Control tab,' select FTDX101D, your Enhanced COM port, default, default, default.

On the 'Rig Options' tab Set the PTT method to RTS and your Standard COM port. In 'Mode,' select 'Data/Pkt.'

In 'Split Operation,' select Rig.

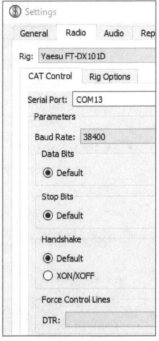

Figure 12: JS8Call setup

> **MMTTY**

MMTTY can be used as a standalone RTTY program and it is often used in conjunction with logging software like DX4Win, or contest logging software such as N1MM.

MMTTY COM port settings

I had a great deal of difficulty getting MMTTY to work with the FTDX101. The answer is to use an addon called EXTFSK v1.06. MMTTY does not use any CAT control, just PTT, the RTTY data signal, and the sound card codec. So, you do not need the Enhanced COM port, just the Standard COM port. Download the correct version of EXTFSK from https://hamsoft.ca/pages/mmtty/ext-fsk.php. At the bottom of the webpage, there is a download link. The download contains some source code files which you **do not need**. Copy only the Extfsk.dll file into the folder containing your MMTTY program files (MMTTYxxxx.exe). That is all you need to do. Then delete ExtFSK106.zip and any remaining extracted files.

Start MMTTY and click the 'Option (O)' tab then select 'Setup MMTTY (O).' Select the TX tab. Click 'Radio Command' and set Port to None. Then click OK.

Set the 'PTT and FSK' dropdown list to EXTFSK (not EXTFSK64). A small popup should appear and it will stay up all the time that MMTTY is working. Note that it has a tendency to hide under other open windows including the MMTTY window. Select your 'Standard' (not enhanced) COM port, FSK = DTR and PTT = RTS. (This must match the settings in the radio).

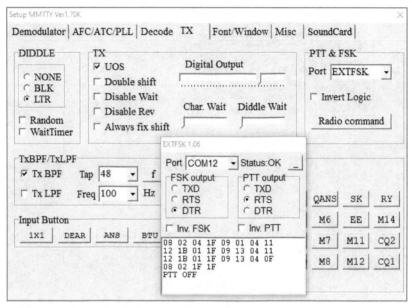

Figure 13: MMTTY and EXTFSK 1.06 config

MMTTY audio settings

Click the 'Option (O)' tab then select 'Setup MMTTY (O).' Select the SoundCard tab.

Set Reception to the FTdx101D RX (USB AUDIO CODEC) and Transmission to FTdx101D TX (USB AUDIO CODEC)

Using MMTTY

Once the COM port has been set up, the EXTFSK popup box will come up every time you start MMTTY and will stay onscreen until you minimize it. It will show the CAT commands as they are sent to the radio. You can minimize it using the button beside 'Status OK,' but you can't close it. It will exit with MMTTY when you shut that down.

Note that you must be in RTTY-L mode on the transceiver because MMTTY is using FSK keying, not AFSK keying like other digital mode programs. This does have the advantage that you can use the onboard message memories and the internal decoder in parallel with MMTTY. But since MMTTY has 16 message memories and its own decoder that "ain't much help."

➤ MixW

No version of MixW will work with Yaesu firmware earlier than V01-20, so make sure that the transceiver's firmware is up to date. I have successfully tested MixW 3.1, MixW 3.2.105, and MixW 4 v1.3.0.

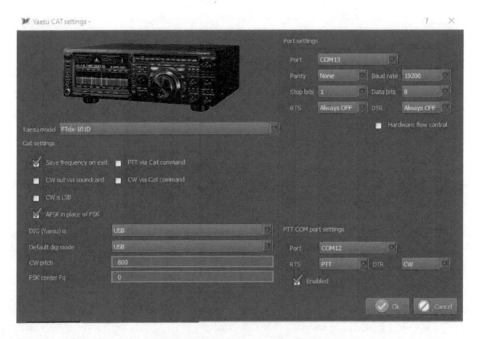

Figure 14: **MixW 4 has a dropdown selection for the FTDX101**

MixW 4 has a dropdown selection for the FTDX101. The CAT settings are selected using the 'Radar screen icon' to the left of the waterfall display. If the icon is not visible click the double arrow icon at the bottom left of the waterfall. Set the CAT up as per the image on the previous page.

The Soundcard settings are selected by clicking the Setup 'cogs icon' at the bottom of the screen, then 'Sound Card.' Initially, I had sound input but no waterfall display. The waterfall settings are found by clicking the double arrow icon at the top right left of the waterfall. You can set the brightness, contrast, and speed.

MixW 3.1 and MixW 3.2 work well with CAT set to FT-5000

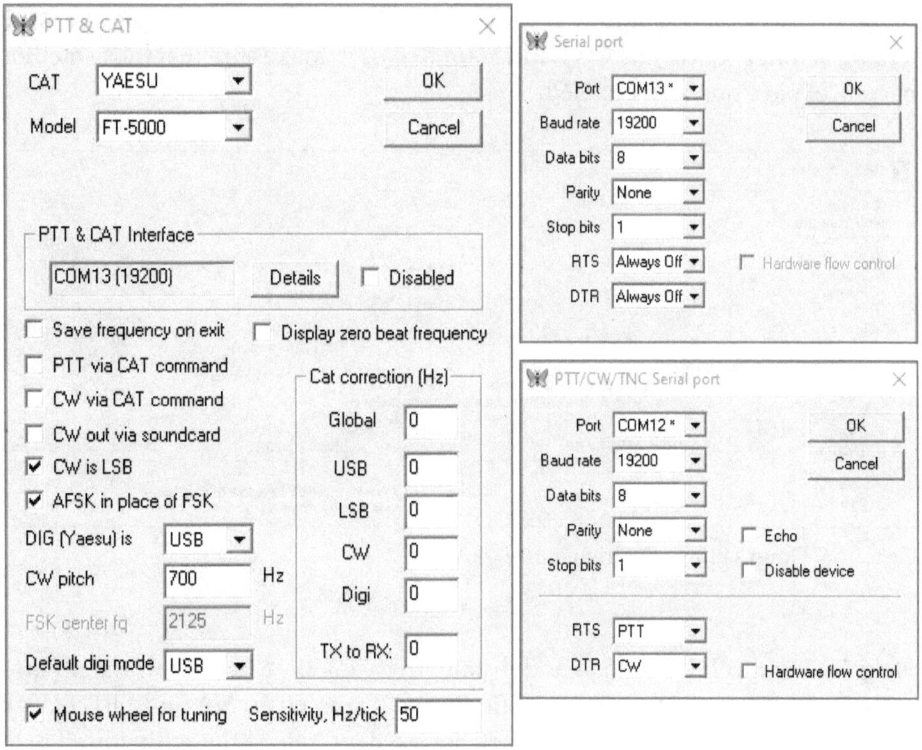

Figure 15: **MixW 3 settings**

Under 'Hardware' '**CAT Settings**' set the Yaesu and FT-5000. In the 'PTT and CAT Interface' box click on 'Details.' Set Port to the Enhanced COM port, Baud rate (same as the radio). I use 19200. Data Bits (8), Parity (None), Stop Bits (1), RTS (Always Off), DTR (Always Off).

Under 'Hardware' '**PTT Port Settings**,' set Port to the Standard COM port, Baud rate can be anything. I use 19200. Data Bits (8), Parity (None), Stop Bits (1), RTS (Always Off), DTR (Always Off). Data Bits (8), Parity (None), Stop Bits (1), RTS (PTT), DTR (CW).

➢ **Fldigi**

Fldigi usually uses HamLib or RigCAT to interface with the transceiver. But the easiest way to get Fldigi to work with the FTDX101 is to download and install Flrig and use that as the radio controller.

Fldigi: Download and install the latest version of Fldigi, I am using V4.1.15. Under 'Configure' – Config Dialog – Rig Setup, check 'Enable flrig xcvr control with Fldigi as client' and 'shutdown flrig with Fldigi.' Leave everything else at the default settings. Check the CAT and Hamlib tabs to make sure that those methods of control have not been enabled.

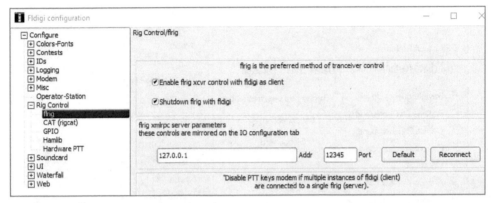

Figure 16: Fldigi setting for Flrig control

Flrig CAT control: Download and install Flrig V1.3.51 or later. Start the program with Fldigi still running. Click Config, Setup, Transceiver. Set rig to FTDX101D, Set Update to the enhanced COM port. Set the Baud rate to the setting used in the radio, 1 stop bit, and RTS/CTS signaling. I left everything else set at the default settings.

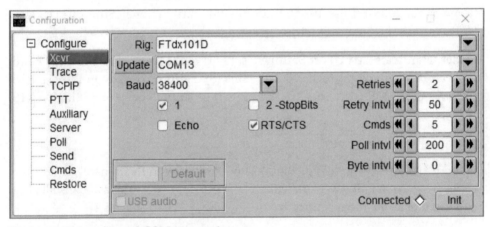

Figure 17: Flrig enhanced COM port settings

Setting up the radio | 51

➢ **Flrig PTT control:** Click Config, Setup, PTT. Under 'PTT control on Separate Serial Port' select the Standard COM port and click PTT via RTS.

➢ **Using Fldigi**

Turn on the transceiver. Start the Fldigi program. Start the Flrig program. At that stage, the frequency display on Fldigi and Flrig should change to match the transceiver's VFO. You can tune using the radio, or Fldigi, or Flrig. There are various other controls on Flrig including volume, notch filter, I.F. shift, attenuator, preamplifier, tuner, and PTT. You can save frequencies to memory channels, change modes, and receiver bandwidth (I.F. width). You must leave the window active, for Fldigi to work but the window can be minimized.

TIP: You can select VFO B, but Fldigi will lose the audio signal. If you want to use VFO B you have to go into the Fldigi configuration and select 'Soundcard, Right channel.' Then check the 'Reverse Left/Right channels' box. When you return to VFO A you will have to uncheck the box again. One way to get around this is to use the A-B button to swap the VFOs over instead of attempting to use VFO B.

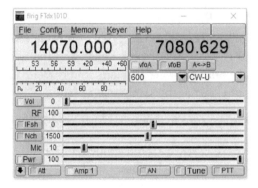

Figure 18: Flrig control software

➢ **Ham Radio Deluxe**

The latest version of HRD supports the FTDX101 and setup is easy. Simply choose the Yaesu and FTDX-101MP options, set the COM port to your Enhanced (CAT) Com port. Set the speed as per the setting in the radio. Finally, check the RTS setting but not the DTR one.

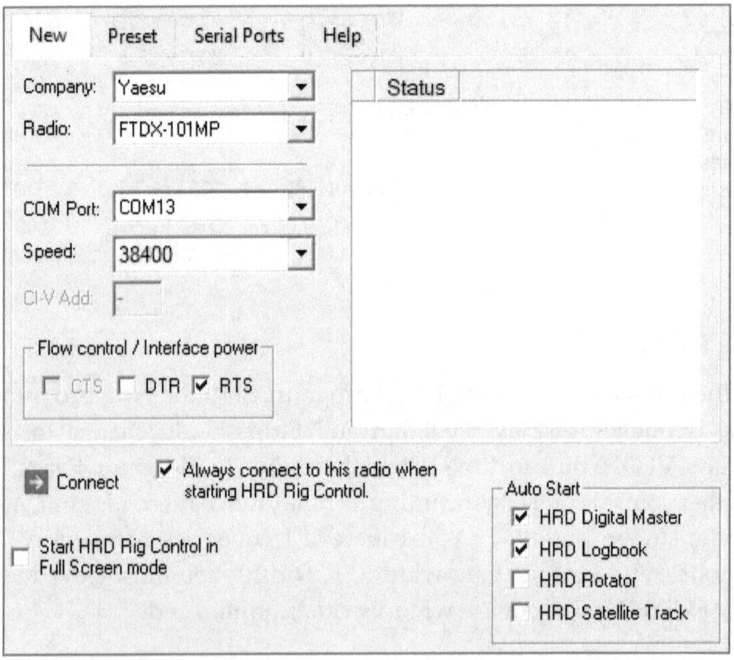

Figure 19: Ham Radio Deluxe setup

➢ **Digital Master (DM780)**

DM780 is usually associated with Ham Radio Deluxe. I had some issues with the application freezing and I was not able to set it up for FSK RTTY. It also locked the transceiver onto transmit, necessitating a power-off reset. However, it works fine with AFSK RTTY and PSK. CW is directly keyed and that works great as well. If you enabled the Auto Start options in Ham Radio Deluxe, DMR 780 and the logging program should start at the same time as HRD.

When you use RTTY make sure that you have selected RTTY-45 AFSK.

Connecting to the Host: If DMR780 is not connecting to HRD, or there are no VFO frequency numbers displayed. Click 'Radio' to show the Radio tab. If the display says 'Closed' instead of showing the transceiver VFO frequency, there should be two buttons above the word 'Closed.' 'Connect to HRD' and 'Configure.'

The 'Configure' tab lets you choose the functions allocated to the buttons and settings below the frequency indicator. Leave the address and Port at 'localhost' and 7809 (or whatever it says). Check the Automatically connect checkbox. Click 'Save' to exit. HRD should connect and display the transceiver's frequency.

You can close the Radio Tab if you want more room on the screen.

TIP: If you want to get back to the configure menu, click the 'play' symbol above the frequency display to disconnect HRD.

The logbook tab: The logbook tab should appear beside the Radio one. If all the squares are greyed out the Logbook is not connected to HRD. Select the 'Program Options' tab and select Logbook. Leave the address and Port at 'localhost' and 7825 (or whatever it says). Check the Automatically connect checkbox. Click 'Save' to exit. HRD should connect and display white squares for log data entry.

PTT settings for CW, digital modes, or voice macros: Click the 'Program Options' tab and select PTT. This should be set for 'via Ham Radio Deluxe …'

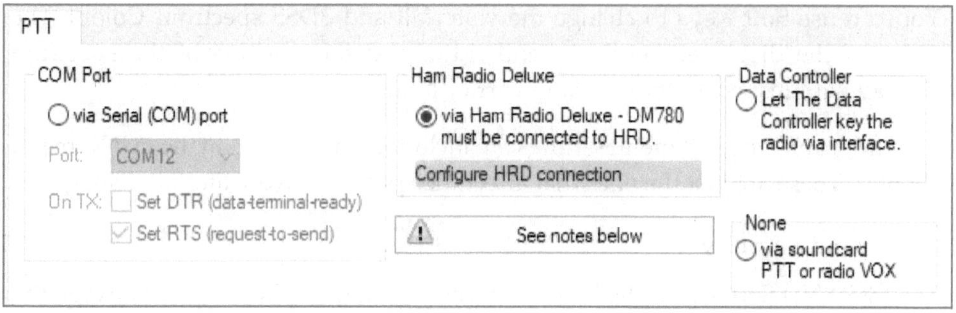

Figure 20: DM780 PTT setting

CW setting: On the Modes and IDs tab, CW sub-tab, you can set the keying signal over the Standard COM port. Check 'Enable serial (COM) port keying' and DTR. Select the Standard COM port number. At the top of the same tab check 'Use PTT.'

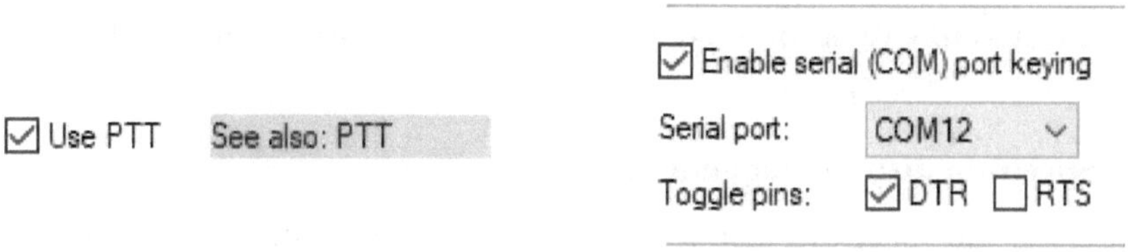

RTTY setting: The FSK RTTY settings are on the RTTY sub-tab. But I was unable to get it to work. Every time I attempted to send FSK RTTY the radio would lock up and an error message said that the program could not find the Standard COM port. It finds just fine on CW mode.

Soundcard settings: Click the 'Program Options' tab and select Soundcard. Select both FTDX101 audio devices. Mine are labeled FTDX101 RX (USB AUDIO CODEC) and FTDX101 RX (USB AUDIO CODEC). It looks like you have to use the Windows level controls to set the audio levels.

> **Using Digital Master (DM780)**

For CW the radio should be in CW mode. This is because the DM780 software is sending CW as a digital signal via the DTR line, not as an audio tone.

For all other digital modes, including AFSK RTTY, the radio should be in DATA-U mode. If you do manage to get FSK RTTY working, set the radio to RTTY-L mode.

SETTING UP THE SPECTRUM AND WATERFALL DISPLAY

You will be using the touchscreen and the spectrum display all the time. This section covers the settings for the spectrum and waterfall display rather than instructions on operating the touchscreen which has a chapter all to itself.

You can use Soft Keys to change the waterfall and 3DSS spectrum Colors, The Color of the displayed receiver bandwidth, and the position of the spectrum display with respect to the oscillator carrier point.

The full set of 'Display' menu settings are included on page 151 in the FUNC menu chapter. These are just the ones that affect the spectrum and waterfall display.

> **Expand Soft Key**

The EXPAND Soft Key at the bottom of the waterfall display toggles between the normal spectrum and waterfall display with a large S meter and filter function display, and the expanded display which has a smaller S meter and a larger spectrum and waterfall display.

> **Waterfall to spectrum display ratio**

Touch the waterfall (not the spectrum display) to change the ratio of spectrum to waterfall on the normal 2D display. There are three options. A small, medium, or large waterfall with a corresponding change in the size of the spectrum display. I prefer the 50/50 display.

> **Spectrum display frequency change**

Touching the spectrum display (not the waterfall) changes the frequency of the radio. It is not very precise unless you are using a narrow span setting, but it does provide a quick way to move the VFO near to a signal you have noticed on the scope. After that, you will have to fine-tune the frequency using the VFO knob.

Setting up the radio | 55

> **Waterfall and 3DSS spectrum Color selection**

Pressing the S.MENU button and selecting the COLOR Soft Key, sets COLOR to the MULTI knob and pops up the 'Color picking dialogue box' for 4 seconds.

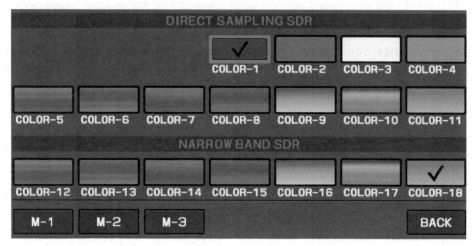

Figure 21: **Colour picking dialogue box**

Alternatively, pressing FUNC and selecting the COLOR Soft Key sets COLOR to the MULTI knob but does not open the 'Color picking dialogue box.' Either way, as soon as you turn the MULTI knob the box will open for 4 seconds, and you can choose from the available Color selections.

The first 11 Color selections affect the wide waterfall or 3DSS display. The part of the display that is generated by the 'direct sampling SDR.' Some of them also change the Color of the spectrum outside of the receiver passband.

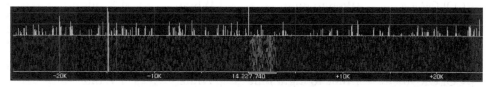

Figure 22: **Colours 1-11 are for the wideband display, 12-18 are for the narrowband section**

Selections 12 to 18 change the waterfall Color of the frequencies that lie within the receiver passband. They are labeled 'Narrow Band SDR' because they are the frequencies that are being sent to the Narrowband SDR which drives the receiver chain. The Colored area is the width of the currently selected roofing filter, not necessarily the I.F. bandwidth. It does not show the effect of using the I.F. width control, but on the FIXED or CURSOR band scope setting it does show the effect of the I.F. shift control.

Color choice: The Colors you choose will be used for both the 3DSS and the normal spectrum and waterfall display. They affect the display on all bands, but you can choose a different Color scheme for the sub-receiver.

TIP: It is not a bad idea to use a different Color scheme for the sub-receiver, so you can see instantly which spectrum scope you are looking at.

Color memories: You can save your three favorite Color combinations by touch and holding M-1, M-2, or M-3 until you hear a double beep. Touch M-1, M-2, or M-3 to recall a saved Color combination.

Color map: These two tables describe the Color combinations.

Color map – wideband area			
Number	Waterfall/3DSS	Signal	Spectrum
COLOR-1	Blue	White	White
COLOR-2	Light blue	White	White
COLOR-3	Grey	White	White
COLOR-4	Amber	White	White
COLOR-5	Red to Blue	White	White
COLOR-6	Blue to Red	White	White
COLOR-7	Violet to Blue	White	White
COLOR-8	Blue to Violet	White	White
COLOR-9	Blue to Light Blue	White	White
COLOR-10	Light Green to Blue	Light Green	Light Green
COLOR-11	Red to Orange	Yellow	Cream

Color map – narrowband area			
Number	Waterfall/3DSS	Signal	Spectrum
COLOR-12	Red to Blue	White	White
COLOR-13	Blue to Red	White	White
COLOR-14	Violet to Blue	White	White
COLOR-15	Blue to Violet	White	White
COLOR-16	Blue to Light Blue	White	White
COLOR-17	Light Green to Blue	Light Green	Light Green
COLOR-18	Red to Orange	Yellow	Cream

> **Dual spectrum and waterfall displays**

Both spectrum and waterfall displays are always available. The receivers do not have to be turned on. Dual spectrum and waterfall or 3DSS images can be arranged side by side or Main above Sub. Make sure that MONO is not selected. It should be white. Press DISP repeatedly to cycle through the four display options.

1. One big spectrum display. Can be 3DSS or normal.
2. Two spectrum displays SUB above MAIN. Either or both can be 3DSS or normal 2D. The touchscreen controls affect the Main or sub receiver according to which receiver is selected by touching the VFO frequency display or by using the MAIN and SUB buttons.
3. Two spectrum displays, with MAIN on the left and SUB on the right. Either or both can be 3DSS or normal 2D. The touchscreen controls affect the Main or sub receiver according to which receiver is selected by touching the VFO frequency display or by using the MAIN and SUB buttons.
4. The spectrum displays are replaced with the big meters and larger versions of the DSP filter function displays.

Modifiers

- The DISP button will not work if the display is in MONO mode.
- Pressing the DISP button deactivates the MULTI display.
- Pressing the DISP button deactivates the HOLD display.
- Touching EXPAND and then using the DISP button cycles, through slightly larger spectrum displays with the usual reduced metering. The last option with no spectrum displays is unaffected by the status of the EXPAND control.

➢ **Spectrum scope menu settings**

RBW (Resolution bandwidth): There are three choices of resolution bandwidth for the spectrum scope. I can't see any reason why you would want to change from the default HIGH setting. <FUNC> <DISPLAY SETTING> <SCOPE> <RBW>.

Carrier point: The easiest way to see the effect of the carrier point control is using the 'CENTER' display mode on SSB.

To change the scope center point, select <FUNC> <DISPLAY SETTING> <SCOPE> <SCOPE CTR> and select CAR POINT or FILTER.

If you choose FILTER, the receiver passband, which is usually be highlighted on the waterfall in a different shade or Color, will be placed right in the center of the screen, with the white center line in the middle of the receiver passband. If you turn on the Marker you will see the real VFO frequency on the left side of the receiver passband for USB, or the right side if the radio is set to LSB.

I strongly recommend using the 'carrier point' (CAR POINT) option which places the real VFO frequency on the white center line and displays the receiver passband on the right for USB, on the left for LSB, or in the center for AM, CW, or FM.

Markers: I recommend that you turn the markers on if you are using any display mode other than CENTER. A vertical line will be shown on the spectrum display at the carrier point frequency. The green marker indicates the receiver VFO frequency, and the red marker indicates the transmitter VFO frequency. In simplex mode, they overlay each other with the red marker on the top. You will be able to see both markers if you select Split operation.

To toggle the markers on or off, select <S. MENU> <MARKER>

If you tune above or below the range of frequencies in the currently displayed span, the markers will sit at the edge of the display and the entire spectrum and waterfall or 3DSS display will scroll as you tune. Touching the spectrum or waterfall will move the VFO to a frequency close to the position you touched.

2D and 3DSS sensitivity: There are 'HI' and 'NORMAL' sensitivity settings for the 3DSS or waterfall and spectrum display. Try both settings and see which you prefer.

<FUNC> <DISPLAY SETTING> <SCOPE> <2D DISP SENSITIVITY>

<FUNC> <DISPLAY SETTING> <SCOPE> <3DSS DISP SENSITIVITY>

I prefer the HIGH setting for both the 3DSS display and the 2D display.

TIP: The 3DSS setting seems to make the display very slightly more sensitive when the HIGH setting is chosen. Which is what you would expect. However, the 2D setting seems to do the exact opposite. The display is much brighter on NORMAL than it is on HIGH.

Waterfall and 3DSS scroll speed: Somebody online suggested that the waterfall speeds are adjustable from "too fast" to "much too fast." Use <S.MENU> <SPEED> and rotate the MULTI knob. The selection stays associated with the MULTI knob so you can experiment and see what you like. I am happy with the SLOW-2 setting.

Waterfall and 3DSS peak: This setting changes the dynamic range of the waterfall display. If you are on a noisy band or there are large signals you might want to reduce the peak level. Use <S.MENU> <PEAK> and rotate the MULTI knob. The selection stays associated with the MULTI knob so you can experiment and see what you like. I usually leave it set to LV4.

Spectrum level: You will be adjusting the spectrum level constantly. It changes every time you change the receiver bandwidth or change bands. There really should be an easy to reach control for it, but there isn't. Use <S.MENU> <LEVEL> and rotate the MULTI knob. The selection stays associated with the MULTI knob until you make another menu selection.

TIP: You can also associate the spectrum level to the big VFO ring by press and holding the CS button and selecting LEVEL. This control is so essential I leave CS set to LEVEL and the CS function turned on all the time.

Setting up the radio | 59

➢ **Adjusting the 2D spectrum level**

Adjust the level for the 2D display until you see a line of 'grass' across the bottom of the spectrum display. At that setting, the waterfall display Color should also be optimal. Note that adjusting the spectrum and waterfall display does not affect the receiver's performance at all.

TIP: If you adjust the grass level but want the waterfall brighter, change the <FUNC> <DISPLAY SETTING> <SCOPE> <2D DISP SENSITIVITY> setting to NORMAL.

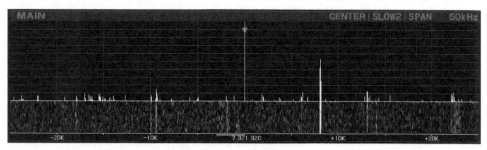

Figure 23: **Adjusting the 2D spectrum and waterfall**

➢ **Adjusting the 3DSS display level**

Adjust the level for the 3DSS display until the background noise is about a 50% mix of black and Colored speckles. That way you maximize the dynamic range of the display. Signals should show as rows of spikes receding into the distance. The super-bright images on the Yaesu advertisements and most websites showing the 3DSS display have the level set much too high. I noticed that the flyer for the new Yaesu FTDX10 has a better picture.

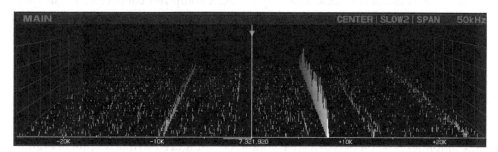

Figure 24: **Adjusting the 3DSS display**

CONNECTING AND USING A LINEAR AMPLIFIER

> Linear amplifier connections

Connect a PL259-PL259 RF cable from the ANT output connector on the radio to the AMP INPUT on a linear amplifier. The linear amplifier output will be connected to the antenna, possibly via an antenna tuner and/or a Power or SWR meter.

If you have a Yaesu VL-1000 linear amplifier it will be controlled via a Yaesu CT-178 cable connected to the 'LINEAR' connector on the rear panel of the radio. This connection takes care of PTT, ALC, and automatic amplifier band switching.

If you have a non-Yaesu linear amplifier you can provide the ALC and PTT connections using two RCA cables from the TX GND connector and EXT ALC connector on the radio. You can use a standard stereo audio cable with RCA connectors. I always use the red connector for PTT and the white for ALC. Do not connect the amplifier PTT to the PTT connector on the radio, as it is an input only.

Or you can create or buy a D-sub 15 pin cable for the LINEAR port and use that for the ALC and PTT connections to the amplifier. I already had RCA cables, so I went with that option.

The TX GND output on the 15 pin, or RCA jack, is an open collector transistor capable of sinking 200 mA at up to 60 volts, or 1 amp at up to 30 volts. It should have no problem switching even ancient linear amplifiers.

Some linear amplifiers are compatible with the Yaesu band data output. Or you could build a band decoder using an Arduino or similar. The connector is a 15 pin D-sub connector sometimes known as a DA-15. It is the old larger type of RS232 connector. It is not the same as a 15 pin SVGA computer connector. Yaesu band data is covered on page 101.

> Linear amplifier ALC

An ALC (automatic level control) connection between the linear amplifier and the transceiver is not essential but it is recommended. The cable goes directly from the ALC output on the amplifier to the ALC input connector on the FTDX101. The ALC output from the linear amplifier turns down the transceiver output power if the amplifier is being overdriven. It should be configured so that it is not operating unless the transceiver is accidentally left at full power when driving the amplifier. Don't use it as a method of controlling the amplifier power. It should only be used as a failsafe in the event of a power setting mistake.

ALC also helps to protect the linear amplifier from 'overshoot.' Some transceivers emit a full power RF spike when you key the transmitter even when the RF power is set to a low level. Without ALC control this tends to trip the protection circuit in the amplifier.

➤ Setting the linear amplifier ALC level

Find out how to adjust the ALC output level on your amplifier. It might be a menu setting as in the case of my Elecraft amplifier or it might be a control on the front or rear panel of the amplifier. Press <FUNC> <RF POWER> on the Yaesu radio and use the MULTI knob to turn the RF Power level down to about 30 watts. If your linear amplifier needs a lot of drive, this could be higher. It is easiest to use CW with the keyer turned off or your digital mode of choice. Do not transmit at full power for an extended period. A minute at a time should be more than enough.

Remember to identify your station as you transmit. Increase the RF POWER level until the Linear Amplifier peaks to full power, (or the highest power that you want to run). Note the setting of the RF POWER control for future reference.

Now, while transmitting, increase the ALC level from the linear amplifier until the output power from the transceiver just starts to decrease. Then back the control off slightly so that full power is being generated and the linear amplifier ALC is not having any effect on the output power. This 'threshold point' is the correct setting. If you increase the RF POWER from the Yaesu transceiver, the linear amplifier should be prevented from going into an overload situation.

The ALC control is an amplifier protection method, like a circuit breaker or a fuse. You can get spurious outputs and intermodulation if you intentionally run full power from the radio and rely on the ALC to limit the RF power. Always reduce the RF LEVEL control to the safe level that you determined above. That way the ALC should never operate, but it is there in case of "finger trouble."

➤ Digital modes

I can't find anything in the Yaesu documentation that suggests that you have to reduce the transmitted RF power for continuous transmission modes like FM, AM, or digital modes. However, it is likely that your linear amplifier is not designed for running at full power for long periods. Check the amplifier manual and if necessary, reduce the output for digital modes. I don't usually run the linear amplifier more than half power on digital modes. In fact, I rarely use the linear amplifier at all. It is not necessary for most contacts. 50 Watts is often more than adequate for modes like FT8 and PSK31.

My Elecraft KPA-500 amplifier is rated for 500 Watts output for up to 10 minutes with a minimum five-minute cool-off period. But I don't find it necessary to run more than 250 Watts on digital modes.

USING AN SD CARD

You can use a full-size SD card to store the radio configuration, memory channel contents, and saved screen capture images, or to upload new firmware. These days most people buy a micro-SD card with an SD adapter.

You can use a 2 Gb SD card or an SDHC card from 4 Gb up to 32 Gb. I am using a 16 Gb SanDisk SDHC card. The information screen says that I have 14.3 Gb free space out of the 14.4 Gb on the card, so buying a 16 Gb card, like I did, is probably overkill.

> Reading and writing to the SD card in your PC

Many laptop computers feature an SD card slot, but it is not common on desktop computers. I purchased a cheap USB to SD card reader so that I can download screen shots for this book and save new firmware updates downloaded from the Yaesu.com website.

> Screen capture

You can take a screenshot picture of the display screen and save it as an 800x480 pixel 'BMP' file on the SD card. Press and hold the FUNC button until you hear a double beep and see 'Screen Shot' on the display. The files will be stored in a 'Capture' subdirectory of the SD card.

> Formatting the SD card

Insert an SD card into the slot below the VOX button. Yaesu says to format it before use, but I didn't bother, and everything works fine.

<FUNC> <EXTENSION SETTING> <SD CARD> <FORMAT>

> Saving and loading memory channels

It is a very good idea to use the SD card to save a backup of your memory channels. They will get deleted if you do a firmware update and you will need the backup to restore them.

Be very careful to use the MEM LIST SAVE command. I know from personal experience that it is very easy to overwrite your memory channels with old data if you accidentally use MEM LIST LOAD.

To save a backup of your memory channel information

Use <FUNC> <EXTENSION SETTING> <SD CARD> <MEM LIST SAVE> <DONE> <NEW> <ENT>

Setting up the radio | 63

To restore memory channel information from a backup

Use <FUNC> <EXTENSION SETTING> <SD CARD> <MEM LIST LOAD> <DONE>. Select the newest file and then click OK when asked to 'Overwrite.' This is your last chance to back out. The radio will reboot after the file has been loaded.

> Saving and loading the radio configuration

It is a very good idea to use the SD card to save a backup of your FUNC menu settings. They may be deleted if you do a firmware update. You will need the backup to restore them.

To save a backup of your menu settings

Use <FUNC> <EXTENSION SETTING> <SD CARD> <MENU SAVE> <DONE> <NEW> <ENT>

To restore your Menu settings from a backup

Use <FUNC> <EXTENSION SETTING> <SD CARD> <MENU LOAD> <DONE>. Select the newest file and then click OK when asked to 'Overwrite.' This is your last chance to back out. The radio will reboot after the file has been loaded.

> Information

<FUNC> <EXTENSION SETTING> <SD CARD> <INFORMATIONS> <DONE> pops up a screen that shows the storage size of your SD card and how much space is available.

FIRMWARE UPDATES

Updating the firmware is always a bit scary because if it goes wrong it can turn your radio into a "brick." However, I found that updating the firmware went very smoothly. Some of the features discussed in this book require the firmware to be version V01-20 or newer.

It is very important to update the radio to the latest firmware release.

Firmware updates since the radio was released have introduced the 'Preset' mode, improved the function of the ANT3 antenna connector to allow it to function as a receive-only antenna, and added spectrum level adjustment to the list that the CS button can allocate to the big VFO ring. There were also changes to the Tuner settings, an increase in the allowable receiver bandwidth on SSB, and a change of function to the AMC feature. Which is now always active.

Do not turn the radio off during a firmware update.

➢ Checking the installed firmware revision

You can determine the currently installed firmware revisions using,

<FUNC> <EXTENSION SETTING> <SOFT VERSION>

Your firmware should be,

- Main CPU: (V01-20 or newer)
- Display CPU: (V01-08 or newer)
- Main DSP: (V01-07 or newer)
- SUB DSP: (V01-07 or newer)
- MAIN SDR (FPGA): (V02-06 or newer)
- SUB SDR (FPGA): (V02-06 or newer)
- AF DSP (V01-00 or newer)

➢ Downloading firmware from the Yaesu website

Visit the Yaesu website at http://www.Yaesu.com. Find the FTDX101D or FTDX101MP under products or by clicking the radio image on the banner.

Select the 'Files' tab. In the 'Amateur Radio \ Software' section look for the 'FTDX101_Firmware_Update Information' file. Open the file and compare the firmware version numbers against the firmware currently loaded on the radio.

If the latest release is newer than the one currently installed on the radio, you should download the new firmware. If the latest release is not newer than the firmware already on the radio. Sit back and relax. You do not need to do a firmware update.

Download the 'FTDX101_Firmware_Update_date' file. It seems to be the same file for both versions of the radio. Clicking the link will download a zip file to your PC. The most recent firmware include in this book is, FTDX101_Firmware update_202104A.zip.

Unzip the files and save them in the FTDX101 directory on the SD card. Do not save them into a sub-directory. If your PC does not have an SD card slot, you can buy a USB SD card reader very cheaply online or from a computer store. I think mine was less than $5.

Insert the SD card back into the SD card slot under the VOX button on the radio.

➢ Before you update the firmware

Before you update the firmware, save your current menu settings and memory channels. They may be overwritten when you update the firmware, and you will want to recover them afterward.

Setting up the radio | 65

To save a backup of your menu settings use, <FUNC> <EXTENSION SETTING> <SD CARD> <MENU SAVE> <DONE> <NEW> <ENT>

To save a backup of your memory channel information use, <FUNC> <EXTENSION SETTING> <SD CARD> <MEM LIST SAVE> <DONE> <NEW> <ENT>

➢ Updating the firmware

Once you have saved the backup files, downloaded the firmware file from Yaesu, saved it on the SD card, and reinstalled the card in the transceiver, you are ready to install the new firmware.

<FUNC> <EXTENSION SETTING> <SD CARD> <FIRMWARE UPDATE> <DONE> pops up a screen which shows the firmware version on the SD card and allows you to update the radio.

You can select which of the seven options you want to update. Normally you would choose to install all the updates. If the downloaded file contains no update for one of the processors, it will indicate '(No such file)' and you can't select that line.

Select the firmware that you want to update by ticking all the check boxes, then touch UPDATE, and then OK.

Do not turn the radio off during a firmware update.

The radio will re-boot after the new firmware has been installed, to load the revised files into the various DSP, FPGA, and CPU devices.

If an error message is shown during the firmware update, for example, a corrupt file has been detected. Touch the screen and the radio will re-boot automatically. Then download a fresh copy of the firmware from Yaesu and repeat the update process.

Operating the radio

The Yaesu is a joy to use. I like the feel of the main-VFO tuning knob. The drag is adjustable, but I like it just the way it came from the factory. Generally, the ergonomics are good although some things take a bit of getting used to. I have assumed that as the purchaser of an elite class transceiver, you will be familiar with the basics of operating an HF transceiver. This chapter includes some differences in the way that you operate this radio compared to other models. For example, since the radio has two completely independent receivers, operating in Split mode is different from what you might expect.

This chapter includes operating split mode, operating split mode without using the Split function, SO2R/SO2V operation, using external digital mode software, computer mouse operation, and using the Scan mode. There is also a section on operating CW which covers the excellent APF filter and the break-in setting which is a little different in the FTDX101. The audio scope display is a great addition. It can show you the audio frequency of an interfering signal and the effect of the manual notch filter, contour control, and VC tune control. In CW mode an optional bar graph shows you the exact tuning of a wanted CW signal.

TIP: You may be wondering why I didn't place the chapters about the touchscreen and the spectrum display before the chapter about operating the radio. I did it for two reasons. Firstly, it is more interesting to read about how you will actually use the radio and secondly because the discussion about the touchscreen naturally leads on to the spectrum display and that in turn to the Soft Keys and the FUNC menu, then the sub-menu options. I believe it is natural to follow that with the front panel controls and that leads on to the rear panel connectors. So overall I believe that it was better to put the Operating the Radio chapter first.

ANTENNA SWITCHING MATRIX

The FTDX101 has three antenna connectors. You might have a triband Yagi connected on ANT1, a low band antenna on ANT2, and perhaps a 6m beam or a receive-only antenna connected to ANT3. In normal 'TRX' operation the radio transmits on the same antenna port as it is using for receiving. But this can be changed if you want to use the ANT3 antenna connector as a receive-only antenna.

<FUNC> <OPERATION SETTING> <GENERAL> <ANT3 SELECT>

TRX	Transmit and receive on ANT3	R3-T1	Transmit on ANT1 and receive on ANT3
R3-T2	Transmit on ANT2 and receive on ANT3	RX-ANT	Receive on ANT3, transmit disabled

Select the antenna that will be used for receiving using the ANT Soft Key on the display. It is above the left side of the scope display, under the VFO MHz display. The last four options are controlled by the ANT3 SELECT setting.

ANT	Operation
1	Transmit and receive on ANT1
2	Transmit and receive on ANT2
3	Transmit and receive on ANT3
R/T1	Transmit on ANT1 and receive on ANT3
R/T2	Transmit on ANT2 and receive on ANT3
RANT	Receive on ANT3. The transmitter is disabled

GENERAL RECEIVER OPERATION

The FTDX101 has two fabulous receivers with some really neat features. I am assuming that you have owned an HF transceiver before and know the basics. This section covers some of the special features that may be different from your previous radio(s).

➢ **Mono mode**

As you are only interested in working on one band at present, touch the MONO Soft Key to turn off the sub-VFO display. The MONO Soft Key will turn blue. Note that this does not turn off the sub receiver or change the transmitter frequency. It only turns off the second display. The DISP button is disabled in the MONO mode because it is only used to change the way the dual receivers are presented on the screen.

➢ **Spectrum scope span**

It is usual to use the 3 kHz roofing filter for SSB. I use a 50 kHz span on the spectrum display. It provides a good display of the band and the representation of the receiver bandwidth is not too big and not too small. A span of 100 kHz looks OK as well. In a contest, you may want to see more of the band so you might choose a 200 kHz or 500 kHz span.

Use a 100 kHz or 200 kHz span for FM or AM because those modes have a wider bandwidth.

You usually use the 600 Hz roofing filter for CW. On the FTDX101MP you can select a 600Hz or a 300 Hz roofing filter. You can tune the marker onto the wanted frequency and use ZIN or the audio bar graph for fine-tuning. Choose a span that covers most of the CW portion of the band. Anything between 20 kHz and 100 kHz looks OK. I often choose a 50 kHz span as it provides a good balance between the range of displayed frequencies and the narrow receiver bandwidth.

Trimming the grass

Press and hold the CS button and select the LEVEL control. Leave CS turned on and use the big VFO ring to reduce the noise on the spectrum scope "grass level" so that it is just showing a millimeter or two (0.1 inches) above the top of the waterfall display. At that point, the waterfall display Colors should look optimal. On the 3DSS display adjust the level a bit lower for a speckled carpet of Color, about 50% black.

Expand

The EXPAND Soft Key creates a larger spectrum and waterfall display, reducing the size of the meters. In Expanded Mode, the meters only show the S meter and the currently selected transmit metering instead of all the meter scales.

Preamp and attenuator

The 'normal' IPO setting is AMP1, with ATT (the attenuator) turned off. Using AMP2 is appropriate on the 10m and 6m bands where there is less band noise. And on the noisy 80m and 16m bands IPO is the best option. Possibly with some attenuation as well if the received signals are strong. Turning off the preamplifiers by selecting IPO and adding some front-end attenuation will improve the dynamic range and as a result the signal to noise ratio of strong signals.

Voice keyer

You can key the voice, CW, or digital modes keyers from the <FUNC> <REC/PLAY> menu popup, or the FH-2 external keypad. You must turn BK-IN on to transmit a voice or CW message on the air. Otherwise, you will hear it, but nobody else will.

The RTTY and PSK keyers do not require you to turn BK-IN on. In fact, the BK-IN button is disabled in the Data, PSK, and RTTY modes. See setup on page 20.

RF gain and squelch

The default setting for the RF/SQL control is for it to act as an RF Gain control. When set to RF, the knob provides a manual adjustment of the gain of the RF and I.F. stages. It can be very useful to turn down the RF Gain to improve the signal to noise ratio of reasonably strong signals, especially on the noisy lower frequency bands.

But most people leave the RF Gain turned up to maximum all the time. The Yaesu manual states that the "*RF/SQ* knob is normally left in the fully clockwise position." I find it much more useful to change the menu setting to SQL so that the control operates the receiver squelch. My recommendation is if you operate mostly on 40m, 80m, and 160m set the menu to the RF gain position. If like me, you favor the higher bands, set the menu to SQL. The menu setting affects both receivers.

<FUNC> <OPERATION SETTING> <GENERAL> <RF/SQL VR> (RF or SQL)

TIP: I removed the RF/SQL knob and replaced it offset by 180 degrees to stop the squelch control from getting in the way of the MULTI knob. It makes the control work backwards, but you quickly get used to that. I did the same with the sub-VFO so that the action is the same for both receivers.

> **Dual receivers**

This radio has two identical receivers, so you might as well use them. Touch the MONO Soft Key so that it is white, representing the dual receiver mode. Touching the VFO frequency display will swap the main and sub-VFOs, indicated by the MAIN and SUB buttons at the top of the VFO tuning knob. But it will not turn on the sub receiver. To do that you have to turn it on with the RX button or the SYNC button. You can listen to two signals on different bands, or the same band. The two receivers can be on different modes. It can be useful to set one receiver to a net or sked frequency, while you tune around looking for interesting DX. It is also very handy to use the sub-receiver to listen to the pileup of stations attempting to work a DXpedition or a rare DX station. You can determine which way the DX station is working through the stations calling, or simply find a quiet spot to make your call.

The sub receiver controls, marked with blue lettering, are above the main receiver controls marked with white lettering. The band indicator has blue LEDs for the sub receiver above the white LEDs which indicate the band for the main receiver.

TIP: Press and holding a band button illuminates an orange LED. This has no function, but it can serve to remind you which bands you can operate on during a contest, or which bands to use with a particular antenna. For example, I turned three on for the bands where I can use my Tri-band beam. However, I find that I have a tendency to turn them on or off by accident if I get too aggressive when I am switching bands.

- The SYNC button locks the VFOs together so that tuning the main channel VFO changes the frequency of the sub receiver and the main receiver at the same time. The tuning rate is as per the main-VFO. Press and hold the SYNC button to set the sub receiver to the same frequency as the main receiver.

- The SPLIT button moves the transmitter to the sub-VFO frequency. But it does not turn on the sub receiver. You need to be very careful because the sub-VFO frequency might be on another band and/or set to a different mode. Holding the SPLIT button until a double beep is heard, sets the sub-VFO to the pre-set split offset and the mode the same as the main-VFO. Holding down the SPLIT button again offsets the transmitter by a second multiple of the pre-set split offset. Each time you press and hold the button the transmit frequency steps up or down the band.

The audio output on the USB cable has the main receiver audio on the left channel and the sub receiver audio on the right channel. The transmitter audio can be on either or both channels.

> **VC Tune**

I like the VC Tune control very much. It is one of the best features of the radio. The VC TUNE button enables a tracking preselector. It is excellent if you have large interfering signals on the band such as you might experience during a contest weekend, or if nearby shortwave stations are creating intermodulation interference. You can manually adjust the VC Tune filter to help to eliminate interference, or just peak the received signal level shown on the S meter or spectrum display and let it automatically track any frequency changes that you make. If you switch bands the radio will remember whether the VC tune was previously switched on for that band and the tuning setting. The beauty of the VC Tune system is that eliminates interference at the RF stage before the bandpass filters and before it gets into the I.F. and the DSP stages of the receiver. As a result, it is effective at reducing intermodulation distortion.

The VC (variable capacitor) tune system uses a filter where the capacitance is changed by a variable capacitor driven by a stepper motor. If there are no interfering signals you should turn VC Tune off.

TIP: The VC Tune control is a fairly broad filter, but if it is tuned slightly off the center of the spectrum display, or you are using a very wide spectrum or 3DSS span you may see some attenuation in the noise level. This is a normal effect of the filter design.

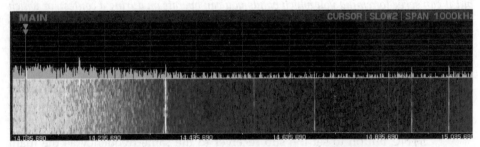

Figure 25: **The VC Tune setting can desensitize the spectrum display**

The FTDX101D has a VC Tune preselector unit on the main receiver. The FTDX101MP has a VC Tune preselector unit on both receivers.

You may have noticed that there are two VC TUNE buttons at the top right of the VFO knob. Pressing the larger one allocates manual control of the VC Tune frequency to the big VFO ring. It also turns on the other VC TUNE button. A red bar graph above the DSP filter function display indicates the frequency of the VC Tune filter. Note that this control is working at the received RF frequency, not the audio signal shown on the filter display below the bar graph.

It is working on the RF signal right at the front-end of the receiver. The smaller VC TUNE button turns the VC Tune function on or off without enabling the VFO ring manual tuning.

Initially after pressing the larger VC TUNE button, use the VFO ring to peak the noise 'grass' level on the spectrum or 3DSS display. This centers the filter on the band. After that, it will track any tuning that you do unless you switch to another band. You can fine-tune either way to reduce an interfering signal. Once that has been done you can press the larger VC TUNE button again to turn off the VFO ring tuning, leaving the filter engaged but no longer adjustable.

TIP: Press and holding either of the VC TUNE buttons resets the capacitor tuning so that the filter is in the center of the currently selected band. (Not necessarily the center of the bar graph). VC Tune does not work on the 5 MHz, 50 MHz (6m), or 70 MHz (4m) bands.

➢ **The receiver clarifier**

The receiver clarifier is equivalent to RIT in earlier radios. It allows you to offset the receiver frequency while leaving the transmitter frequency the same as indicated on the VFO display. It is useful if you are chatting with a station that is slightly off-frequency. Rather than move the VFO which will cause the other station to adjust its frequency the next time you transmit, you can use the receiver clarifier to fine-tune the incoming signal. It is especially useful in a net where most stations are on frequency, but one is a little off frequency.

Pressing the Clarifier RX button (not either of the RX buttons above the VFO) enables the RX Clarifier and turns on the CLAR button next to the big tuning ring. You can adjust the receiver offset using the tuning ring. On a FIX or CURSOR display, the green marker will move to indicate the revised receiver frequency. On the CENTER display, the receiver frequency will remain in the center. The red transmitter marker and the VFO numbers will change as you adjust the receiver offset. If you have selected to receive on the sub-VFO frequency by pressing SUB or touching the sub-VFO display, the RX Clarifier will adjust that offset. The sub-VFO RX Clarifier setting is independent of the RX Clarifier setting for the Main-VFO.

Note: If you turn off CLAR, you can still enable or disable the last offset by pressing the Clarifier RX button. You just can't adjust the offset anymore. If you turn off the Clarifier RX button but leave CLAR turned on you can adjust the offset, but it will not affect the receiver frequency until you turn the Clarifier RX button on again.

MORE RECEIVER CONTROLS

The remaining controls are arranged in a group on the right side of the radio. The sub receiver controls, labeled with blue lettering, are identical to the main receiver controls.

➤ I.F. Shift & Width

The WIDTH control changes the bandwidth of the 24 kHz second I.F. inside the DSP stage of the receiver. On SSB it is adjustable from 300 Hz to 4 kHz.

The default is 3 kHz to match the bandwidth of the roofing filter. On CW, PSK, RTTY, and the data modes it is adjustable from 50 Hz to 3 kHz. The default is 600 Hz to match the bandwidth of the roofing filter. The DSP bandwidth is fixed at 9 kHz on AM and 16 kHz on FM.

The current DSP channel bandwidth is shown graphically on the DSP filter function display above the VFO frequency display. The red lines indicate the shape of the filter response. Selecting the 'Soft' 6 dB per Octave setting does not change the red line filter shape, but a blue dip (or peak) indicates the use of the contour control. There are also indicators for the manual notch filter, RTTY mark and space, and the CW APF filter. If you adjust the WIDTH control a popup shows the bandwidth as you make the adjustment. The WIDTH control does not affect the receiver bandwidth depicted on the waterfall display because that shows the roofing filter bandwidth, not the IF bandwidth, but it does change the red lines on the DSP filter function display.

The SHIFT control moves the 24 kHz DSP filter passband by up to ±1.2 kHz to give you some ability to reject interference that is close to the wanted signal and inside the DSP passband. Shifting the DSP I.F. is indicated on both the waterfall and the DSP filter function display. A popup displays the amount of shift you have selected as you make adjustments.

Push and holding the SHIFT knob resets both the I.F. Shift and the I.F. Width functions.

TIP: Pushing the SHIFT knob sometimes sounds the two beeps which indicate that the function does nothing. To clear the shift and width settings, you have to push and hold down the knob until you hear two lower pitch beeps. Perhaps Yaesu could have used the short push to reset the I.F. shift and the long push to reset the I.F. width.

If you are still experiencing interference within the DSP bandwidth you can employ the digital notch filter, the manual notch filter, and the contour control.

➤ DNF digital notch filter

The automatic digital notch filter is effective at removing fixed carrier signals that are creating a beat note on SSB. It can remove the effect of multiple interfering carrier signals that are within the audio bandwidth. To test it, tune the radio so that you can hear a stray carrier signal. Turn on the DNF while looking at the audio filter response, and you will see the filter remove the offending audio signal.

However, you will still see the offending signal on the spectrum and waterfall display because that is showing the incoming RF signal, well before the digital notch filter in the DSP stage.

➤ NOTCH manual notch filter

Turning the NOTCH knob automatically turns on the NOTCH filter button. A red V indicating the notch appears on the DSP filter function display.

Turning the knob changes the notch frequency and this is depicted by moving the slot across the audio spectrum and on a popup display.

I normally tune the notch to the right spot by listening to the interfering signal disappear. Once you have adjusted the notch frequency you can toggle the filter on and off using the NOTCH button. Press and holding the NOTCH knob turns the filter off and resets it to the center of the filter passband.

You can change the width of the manual notch filter. The narrow setting will be fine for single carrier interference and it has less impact on the quality of the wanted signal. The default wide setting will be better at nulling out carriers that are modulated with noise or narrowband data.

<FUNC> <OPERATION SETTING> <RX DSP> <IF NOTCH WIDTH>

TIP: If the interfering signal is very strong Yaesu recommends using the manual filter before trying the DNF notch filter. This is because the manual notch is deeper than the auto notch.

➤ CONT

The contour control was new to me and I quite like it. It produces a small dip or boost in the audio frequency response which you can move across the audio passband from 50 to 3200 Hz. It's like a 'mini-notch.' You can hear the effect as you tune across the audio signal. It can be used as a sort of tone control, attenuating or boosting the low or high audio frequencies. This can be useful if a received signal is "rumbly" or "hissy." It can also attenuate an interfering signal without notching out the audio frequency completely with the manual notch.

Turning the CONT knob (outside ring of the NOTCH knob) automatically turns on the CONT button. A blue 'dip' or 'bump' indicator is shown on the DSP filter function display.

Once you have adjusted the contour frequency you can toggle the filter on and off using the CONT button. Press and holding the NOTCH knob turns the contour filter off and resets it to the center of the filter passband.

TIP: Contour adjustment of the higher audio frequencies has little effect on CW because the 600 Hz roofing filter attenuates those frequencies anyway.

TIP: *If you have a hearing impairment, for example, hearing loss at high frequencies you could set the contour level to a positive value and use the contour function to boost the high audio frequencies. There are no other tone controls so this might be the best solution.*

You can change the bandwidth and the gain of the contour filter. You can even make it a hump so that it acts even more like a tone control, amplifying a part of the audio bandwidth. As an experiment, I set the control to +18 and you could clearly see the increase in noise level as I tuned the contour control across the audio band.

<FUNC> <OPERATION SETTING> <RX DSP> <CONTOUR LEVEL> (-40 to +20)

<FUNC> <OPERATION SETTING> <RX DSP> <CONTOUR WIDTH> (1 to 11)

> **APF**

The APF (audio peak filter) button is for CW only. It is disabled on all other modes.

Tune in a CW signal by ear or using ZIN or the audio bar-graph indicator and turn on APF. The very narrow audio filter eliminates nearly all the band noise revealing a clean CW signal. "It is fab!"

TIP: *The APF filter is very narrow. Unless you know the exact frequency, it is best to turn off APF while you are tuning across the band and turn on the APF after you have the signal tuned in.*

The APF filter frequency is adjustable. Turning the CONT/APF knob while in the CW mode will activate the APF button and enable the knob. APF is adjustable over a range from -250 Hz to + 250 Hz. Press and hold NOTCH to reset the offset to zero.

You can also change the bandwidth of the APF filter. Using the narrow filter setting will provide the cleanest sounding CW signal but the signal has to be exactly on frequency, or you will hear nothing. The wide setting is easier to tune, but you will hear more band noise. To be honest, although I can hear a small difference, I don't think it matters which setting you choose.

<FUNC> <OPERATION SETTING> <RX DSP> <APF WIDTH>

> **NB noise blanker**

Pressing the NB button turns on the noise blanker. Holding the NB button down enables a popup display, and you can use the MULTI knob to adjust the noise blanker level.

You can also achieve this using <FUNC> <NB LEVEL>. Setting the NB level to an aggressive level may affect audio quality, so you may need to experiment with the level when tackling a particular noise problem.

The noise blanker is designed to reduce or eliminate regular pulse-type noise such as car ignition noise. Most DSP noise blankers work by eliminating or modifying noise peaks that are above the average received signal level. They usually have no effect on noise pulses that are below the average speech level.

Note: although you can press and hold either NB button to adjust the noise blanker level, the level setting is shared between the two receivers. Changing the level on one receiver will also change the level on the other receiver. This is a bit odd considering one receiver may be on a band experiencing a lot of impulse noise and the other may be on another band experiencing no impulse noise.

You can also adjust the width and the amount of noise attenuation.

<FUNC> <OPERATION SETTING> <GENERAL> <NB WIDTH> (1, 3, or 10 ms)

<FUNC> <OPERATION SETTING> <GENERAL> <NB REJECTION> (10, 30, 40 dB)

The NB WIDTH control should be set to a short time constant for impulse noise such as vehicle ignition or electric fence noise. You would use a wide setting for longer noise pulses such as lightning crashes or electric welding noise. Noise blankers are usually implemented very early in the DSP process before any DSP filters or demodulation occurs. They "look ahead" in the digital data streams and detect signals that are of short duration and significantly larger than the average speech level. The noise blanker algorithm eliminates or attenuates the noise pulse.

In some radios the DSP noise blanker mutes the noise pulses completely, other developers believe that the receiver sounds more natural if the level in the data sample is reduced to the average of the speech level or attenuated by a fixed value. In this radio, you have a choice of how much NB REJECTION attenuation is applied to data samples where a noise spike has been detected. The default is 30 dB, but you can select a less aggressive 10 dB or a more aggressive 40 dB.

The NB LEVEL adjusted with the MULTI knob, after making a menu selection or by press and holding the NB button, affects the noise blanker threshold. If you use a low threshold the noise blanker will see speech peaks as noise and will attenuate them making the received audio sound distorted. On the other hand, a high threshold setting may mean that the noise blanker fails to react to noise pulses that are only just above the average speech level. It's a trade-off. You will have to make a judgment call on the best setting for a particular problem. Sometimes slightly distorted audio is better than putting up with distracting noise pulses. Start at a low threshold level and increase it until you reach an optimum setting.

> **DNR digital noise reduction**

Digital noise reduction is mostly aimed at beating atmospheric wideband noise. It is more useful on the noisy lower bands.

It appears that the DNR gets more aggressive as you turn the control down from 15. The Yaesu manual states that there are "15 different noise-reduction algorithms" which suggests that the control is not just a linear progression and that some of the intermediate settings may work better than the high or low options.

Aggressive levels of digital noise reduction add an electronic synthesizer 'Dalek' sound to the speech but also works harder to eliminate the background noise. Which do you prefer, and which is less tiring to listen to?

TIP: Digital noise reduction takes quite a long time to calculate the noise characteristic. It can take a second or two to settle into a stable state.

This filter must be important to Yaesu because they have provided four different ways to set the DNR level. You can allocate DNR level to the big VFO ring by press and holding the CS button and selecting DNR LEVEL. You can allocate DNR level to the MULTI knob using <FUNC> <DNR LEVEL>. You can press and hold the DNR button and use MULTI to set the DNR level. Or finally, you can set it in the main menu using <FUNC> <OPERATION SETTING> <RX DSP> <DNR LEVEL>. This setting will not stick if you subsequently use any of the other methods to change the level. In all cases, the range is from 1 to 15 with a default setting of 1. If the DNR level is set to the VFO ring or the MULTI knob, adjusting the level will automatically turn on the DNR filter.

I find that any setting under 5 can cause audio distortion where the received audio spills out past the normal audio filter passband.

➢ DNF, NB, and DNR compared

There is a reason that the buttons for NB, DNR, and DNF are grouped together. They form your toolkit for managing different types of interfering signals. The **digital notch filter** (DNF) works well with fixed signal interference such as carrier signals or ADSL interference. The filter looks at the time constant of the incoming signal and rejects signals that have a long time-constant compared to the ever-changing speech. It is useless for combatting impulse noise but great for eliminating stray carrier signals known as "birdies." The DNF can respond to multiple interference signals within the receiver passband. You cannot use it when you are receiving CW because it will filter out the CW signal that you are listening to.

The **noise blanker (NB)** uses similar DSP code, but different time constant variables. This time the short time constant pulse noise signals are rejected, and the more long-term coherent speech is retained. The noise blanker is designed to reduce or eliminate regular pulse-type noise such as car ignition or electric fence noise. The DSP process looks ahead in the data stream and attenuates the noise spikes with a pre-set amount of attenuation.

Digital noise reduction (DNR) works to reduce atmospheric and other long-term background noise levels, particularly on the noisy lower bands, 40m, 80m, and 160m.

It rejects signals with a very long time constant relative to normal speech or CW signals.

➢ **Markers**

The green marker indicates the receiver frequency, and the red marker indicates the transmitter frequency. The green receiver marker is turned off if the spectrum display is set to CENTER since the receiver frequency is always in the center of the spectrum display. During normal simplex operation the red marker sits on top of the green marker, so you normally only see the green marker if you are operating split in the FIX or CURSOR screen mode.

A red or green arrow at the left or right side of the spectrum scope indicates that the transmitter or receiver frequency is off the edge of the currently displayed spectrum display. This does not affect the operation of the radio, but it means you can't see the relevant frequencies. I recommend leaving the markers turned on all the time unless you always use the CENTER display, and you never use split. <S.MENU> <MARKER> or <FUNC> <MARKER> toggles the markers on and off.

TRANSMITTING

There aren't many mysteries when it comes to operating the transmitter. Everything is much the same as using any other HF transmitter. I am assuming that you have set all the audio levels and CW settings, as discussed in the earlier sections. Other than the mode controls which affect both receiving and transmitting, only ten of the front panel controls directly affect transmitter operation. They are BK-IN, TUNE, MOX, VOX, MONI, CS, the speech processor, the main and sub transmit buttons (TX), and the transmitter clarifier (CLAR).

The biggest risk is inadvertently transmitting on the wrong band. You must be careful because this is fairly easy to do. Above the VFO knob, there are two buttons marked TX. One will have a red LED active and the other will be dark. No matter which receiver is in use. If the left TX LED is lit the transmitter will transmit on the frequency depicted on the main-VFO display. If the right-side TX LED is lit the transmitter will transmit on the sub-VFO frequency.

TIP: The operation of the AMC (automatic microphone gain control) has changed from the original firmware and this has caused some confusion in online forums and the interpretation of the Yaesu manual. Note that contrary to popular opinion, the AMC is active all the time. You cannot turn it off. Pressing the MIC/SPEED knob while in SSB mode will turn on the speech processor (compressor).

TIP: If you are operating with the menu setting that allows CW transmission while in the SSB mode, you will not be able to use the MIC/SPEED knob to turn on the speech compressor. You have to turn the CW in SSB function off, then turn the compressor on, and then turn the CW in SSB function back on again. See 'Troubleshooting' page 182.

➢ Break-in

To send CW the BK-IN button must be activated. It must also be activated if you want to send any of the five voice keyer recordings. In both cases, if BK-IN is not turned on, you will hear the signal, but it will not be transmitted.

You can select SEMI or FULL break-in using <FUNC> <CW SETTING> <MODE CW> <CW BK-IN TYPE>.

➢ Antenna tuner

Pressing the TUNE button activates the radio's internal tuner. It will automatically select the capacitors and inductor that will provide a near match at the current operating frequency. The radio stores the tuner settings after tuning and retrieves them the next time you use a frequency within the same 10 kHz window.

The antenna tuner will not automatically initiate a 'tune' operation if the SWR is high. You have to manually start a 'tune' by press and holding the TUNE button.

The tuner can match load impedances between 16.5 ohms and 150 ohms, (up to 3:1 SWR) on the HF bands, or from 25 to 100 ohms (up to 2:1 SWR) on the 6m band. This means that it should be able to tune out to the band edges of 'resonant' antennas, but it might not be able to tune antennas like G5RV, large loops, long wire, or 42 m verticals. This performance is typical of the antenna tuners in most amateur radio HF transceivers.

There probably isn't enough room in the transceiver for a better tuner. Covering a wider SWR range would require bigger inductor cores and higher voltage capacitors, because of the higher voltages and currents involved with matching extreme loads. Additional heat dissipation could also be a problem.

Holding down the TUNE button initiates an automatic tuning sequence. You will hear all manner of clicks until the radio arrives at the best tuning position. It seems that holding down the TUNE button always initiates a full tune, not starting from any previously saved setting.

➢ External antenna tuner

Th 8 pin TUNER jack on the rear panel is for connection to the Yaesu FC-40 external tuner or a compatible alternative. You should not attempt to use the internal tuner and an external antenna tuner at the same time. Turn off the internal tuner with the TUNE button and tell the radio which antenna port has the external Yaesu compatible antenna tuner. EXT1, EXT2, or EXT3.

<FUNC> <OPERATION SETTING> <GENERAL> <TUNER SELECT>

➢ VOX

VOX stands for 'voice operated switch.' It is available for the speech and data modes. When VOX is turned on, the radio will transmit when you talk into the microphone without you having to press the PTT button. The function is popular if you are using a headset or a desk microphone. Some people always use it. I have never used it.

Turn VOX on using the VOX button. An amber LED indicates VOX is active.

The three VOX control settings are available directly from the FUNC menu. Turn the MULTI knob to change the setting. Or they can be allocated to the big VFO ring by press and holding the CS button. You should take the time to set the VOX up carefully as some settings tend to counteract other settings.

- <FUNC> <VOX GAIN> sets the sensitivity of the VOX, i.e. how loud you have to talk to put the radio into transmit mode. (Default 10).
- <FUNC> <VOX DELAY> sets the pause time before the radio reverts to receive mode. It needs to be set so that the radio keeps transmitting while you are talking normally but returns to receiving in a reasonable time after you have finished talking. (Default 100 milliseconds).
- <FUNC> <ANTI-VOX> stops the VOX triggering on miscellaneous noise like audio from the speaker or background noise. Higher values make the VOX less likely to trigger. The default is zero i.e., no anti-VOX.

TIP: You can get a feedback loop where VOX is triggered by speech from the radio's speaker if you are using the transmit monitor, MONI. The easiest solution is to turn off the transmit monitor. Or you could try turning down the monitor volume and increasing the anti-VOX a bit. Or better still, wear headphones while transmitting so that there is no feedback path to the microphone.

The 4th dimension. You can use VOX with digital mode transmissions rather than messing around with the RTS and DTR control line signals. I would only rely on this method of creating a PTT signal as a last resort if you cannot get standard RTS/DTR COM port signaling to work.

There is a fourth VOX setting deeper in the menu structure that sets the threshold level for 'Data VOX.' The range is 0 to 100 and the default is 50.

<FUNC> <OPERATION SETTING> <TX GENERAL> <DATA VOX GAIN>

➢ MOX

MOX stands for 'manually operated switch.' The switch works in parallel with the microphone PTT switch. It makes the transceiver switch to transmitting in exactly the same way as pressing the mic switch. The radio will keep transmitting until you turn MOX off. A red LED on the button indicates that MOX is active.

Note that in CW mode MOX will allow you to send Morse code from a CW message, or the Morse key or paddle, without having BK-IN turned on.

The microphone is live in the voice modes, so you will transmit any conversations, phone calls, talking to the cat, praising the dog, background music, or audio from another radio. The voice keyer will operate while the radio is being keyed via MOX or the microphone, but as usual, BK-IN must be turned on.

The FUNC button is inactive while the radio is transmitting. Although the screenshot mode still works.

➢ MONI

MONI is the transmit monitor. It allows you to hear the signal that you are transmitting. If you are operating in SSB or AM mode, it is easier to hear and less distracting if you use headphones. You will hear the voice keyer audio, even if the transmit monitor is turned off.

The audio for the transmit monitor is taken from the transmitter I.F. after the modulator but before up-conversion to the final transmit frequency. This makes it especially useful for evaluating the quality and tone of the signal that you are transmitting. Use it when you adjust the parametric equalizer and speech processor (compressor) settings.

You must have MONI turned on to hear the CW sidetone. The transmit monitor is not available in the FM modes.

Press and hold the MONI button and use the MULTI knob to adjust the transmit monitor level. The level is indicated on a popup, but it is better to just adjust it until you are happy with the volume. Alternatively, you can use <FUNC> <MONI LEVEL> to allocate the monitor level to the MULTI knob. When you are in CW mode the sidetone level is adjusted with the same control. Changing the CW sidetone level does not affect the transmit monitor level in the voice modes, and vice versa.

➢ The main and sub transmit buttons (TX)

Above the VFO knob, there are two buttons marked TX. One will have a red LED active and the other will be dark. No matter which receiver is in use and even if the transceiver is in MONO mode with only the main-VFO display active.

If the left TX LED is lit the transmitter will transmit on the frequency depicted on the main-VFO display.

If the right-side TX LED is lit the transmitter will transmit on the sub-VFO frequency. This could cause you to transmit on a different band and/or mode.

Operating the radio | 81

➢ The transmitter clarifier

The transmitter clarifier is equivalent to XIT in earlier radios. It allows you to offset the transmitter frequency from the receiver frequency. It can be another way of operating Split without using the split button. If the right-side TX button is selected and the sub-VFO is activated by touching the VFO frequency display or pressing the SUB button, the transmitter clarifier will move the frequency of the transmitter relative to the displayed sub-VFO frequency.

Pressing the Clarifier TX button (not either of the TX buttons above the VFO) enables the TX Clarifier and turns on the CLAR button next to the big tuning ring. You adjust the transmitter offset using the tuning ring. The red marker will move to indicate the revised transmitter frequency. If you have selected to transmit on the sub-VFO frequency the TX Clarifier will adjust that offset.

Note: If you turn off CLAR, you can still enable or disable the last offset by pressing the Clarifier TX button. You just can't adjust the offset anymore. If you turn off the Clarifier TX button but leave CLAR turned on you can adjust the offset, but it will not affect the transmitter frequency until you turn the Clarifier TX button on again.

➢ The speech processor

The speech processor is only available in SSB. It increases the average power of your transmission by amplifying the quieter sounds while ensuring that louder sounds do not overmodulate the transmitter. In other words, it reduces (or 'compresses') the dynamic range of the modulating audio. The result is more talk power but less natural-sounding speech.

Generally, you would not use speech compression for a local net because the quality of your transmitted audio is important, and your signal is strong. You would use speech compression if you were working a contest or trying to break a 'pile-up' to work a DX station. In that situation, a higher average power outweighs the need for high-fidelity audio. That said, too much compression can make your voice sound distorted and hard to copy, which will actually reduce your chances of working the rare DX.

Press the MIC/SPEED knob to turn the speech processor on or off. An orange LED indicates that 'PROC' is active. You should have already set the speech processor (compressor) level when you set the SSB levels back on page 10. But you can make adjustments using the PROC/PITCH outer control knob. A popup display will indicate the 'PROC LEVEL.'

You can monitor the effect of introducing the speech compressor by listening to your transmitted signal using the transmit monitor (MONI), or by enabling the audio scope and spectrum display using the MULTI Soft Key.

> **CS**

CS stands for 'custom select.' It allows you to allocate one of twelve different functions to the big VFO tuning ring. Since there is no control for adjusting the reference level of the spectrum scope, I usually allocate LEVEL to the tuning ring. Once allocated, pressing the CS button activates the function. If you use the clarifier, VC tune, or Main/Sub buttons, CS will turn off and the selected function will be allocated to the tuning ring. Pressing CS will reactivate the custom selection.

Press and hold CS to choose one of the 12 functions.

RF Power, MONI level, DNR level, NB level, VOX gain, VOX delay, Anti-VOX, Step Dial, MEM CH, GROUP, R.FIL, or LEVEL.

I cannot see any reason to set CS to R.FIL since it is always available by touching the screen. MEM CH is the same as allocating the memory channel to the MULTI knob. You must be in the 'memory channel mode' (V/M) for it to work.

OPERATING IN SSB MODE

By now you will have adjusted the microphone gain, the AMC gain, and the speech processor (compressor) levels. And we have covered the voice message keyer, VOX, MOX, speech processor, transmit monitor, receiver, and transmitter clarifier. It is time to listen to some signals and maybe get brave enough to call CQ.

> **Select SSB**

Select the SSB mode using the SSB button. Repeated pressing toggles the radio between upper sideband, (USB) and lower sideband, (LSB).

Select the CURSOR or CENTER display mode on the 2D spectrum display, or the 3DSS display. If the markers are turned on, you should see a red marker with a shaded or Colored section to the right for USB or left for LSB. The shaded area represents the receiver passband. The rest of the display shows other signals on the band. Tune in an SSB signal by ear or by tuning until the marker is at the edge of the SSB signal on the panadapter with the SSB modulation within the shaded area of the waterfall. Hard to explain, easy to do.

The receiver controls and filters have been covered already. It is normal to set the FTDX101D transmitter for 100 Watts. You can run the FTDX101MP at 200 Watts. You would normally use the speech processor if you were working DX stations or in a contest. You probably would not use it if chatting to a friend or on a Net.

> **ATT, IPO, and R.FIL**

The IPO (preamplifier setting) has already been covered, but to recap, the 'normal' IPO setting is AMP1, with ATT (the attenuator) turned off. Using AMP2 is appropriate on the 10m and 6m bands where there is less band noise.

And on the noisy 80m and 16m bands IPO is the best option. If you need to use the ATT attenuator, turn the preamplifier off (to IPO) first. Otherwise, the preamplifier is adding gain, partly negating the effect of the attenuator. The receive filter R.FIL should be set to 3 kHz for SSB.

➤ The 2D and 3DSS display level

If you are using the 2D display, adjust the level so that the noise level is just showing a little 'grass' at the bottom of the spectrum display. I allocate the Spectrum Level to the big VFO ring using the CS function. At that level, the waterfall should have a good Color, not washed out, with good contrast where there are signals.

If you are using the 3DSS display, adjust the level so that the noise level is a mottled carpet about 50% black. I don't like it to show any 'grass,' just a newly mown lawn. That way the signals jump out with maximum dynamic range and good contrast.

OPERATING IN CW MODE

➤ Break-in setting

The transceiver will not automatically transmit CW unless BK-IN has been selected using the BK-IN button. The radio will operate in full Break-in or Semi break-in depending on the menu setting.

<FUNC> <CW SETTING> <MODE CW> <CW BK-IN TYPE> (SEMI or FULL)

If BK-IN is set to OFF the transceiver can be made to transmit by pressing the MOX button, pressing the PTT button on the microphone, sending a CAT command, or grounding the SEND line on either of the ACC jacks.

The break-in setting affects the sending of keying macro messages and Morse Code send from a key or paddle, but not CW sent from an external computer program.

- With BK-IN OFF you can practice CW by listening to the side-tone without transmitting.
- Full break-in mode BK-IN FULL will key the transmitter while the CW is being sent and will return to receive as soon as the key is released. This allows for reception of a signal between CW characters.
- Semi break-in mode BK-IN SEMI will key the transmitter while the CW is being sent and will return to receive after a delay when the key is released. The Semi Break-In delay is set using <FUNC> <CW SETTING> <MODE CW> <CW BK_IN DELAY>. It is adjustable from 30 ms to 3 seconds. The default is 200 ms.

➢ Sidetone

The CW sidetone is only heard if the transmit monitor (MONI) is turned on. Press and hold MONI or select <FUNC> <MONI LEVEL> to adjust the monitor level with the MULTI knob. This adjustment does not affect the transmit monitor level on SSB.

➢ ZIN/SPOT

ZIN is an auto-tune feature that pulls the receiver frequency onto a CW signal by matching the CW note with the Keyer pitch, netting it to the transmit frequency.

TIP: The manual reference on page 33 is misleading. It says to press the SELECT switch.

While receiving a CW signal, press ZIN/SPOT to pull the receiver VFO so that the received tone is the same as the keyer pitch. At that point, the transmitter will send on exactly the same frequency as you are receiving.

Most of the time it works very well, although you may have to press the button two or three times to get exactly on frequency. But it does not always work. Sometimes pressing the ZIN/SPOT switch can sometimes move you in steps away from the wanted frequency.

If you press and hold the ZIN/SPOT button the keyer pitch tone will be heard, so you can compare the tone of the incoming signal with the keyer pitch. You can tune the VFO and 'zero beat' the audio from the two sources.

➢ APF filter

The APF button is only active in the CW mode. It enables the 'Audio Peak Filter' which is very effective at lifting weak CW signals out of the noise or eliminating other CW signals that are very close to the receiving frequency. It is best used in conjunction with the ZIN/SPOT button. Use ZIN to pull the radio onto the exact frequency being used by the transmitting station and then use APF to eliminate everything except that station.

When the filter is engaged a red line indicating the APF frequency is displayed on the DSP filter function display.

The APF filter frequency is adjustable over a range from -250 Hz to + 250 Hz. Turning the CONT/APF knob while in the CW mode will activate the APF button and enable the knob. Press and hold the NOTCH knob to reset the offset to zero.

TIP: Unless you are on a Net with multiple CW operators, I think it is more likely you will leave the APF at the default center frequency and adjust the radio VFO so that the CW signal is centered on the filter.

You can also change the bandwidth of the APF filter. Using the narrow filter setting will provide the cleanest sounding CW signal but the signal has to be exactly on frequency, or you will hear nothing.

The wide setting is easier to tune, but you will hear more band noise. To be honest, although I can hear a small difference, I don't think it matters which setting you choose. <FUNC> <OPERATION SETTING> <RX DSP> <APF WIDTH>

> **Key speed and pitch**

The PROC/PITCH control sets the speed of the electronic keyer including CW sent from the message memories. As you adjust the control the CW speed in WPM (words per minute) is displayed on the usual popup. Pressing the PROC/PITCH knob turns the electronic keyer on or off.

The outer PITCH control changes the pitch of a received CW signal without changing the receiver frequency. You can set the control so that CW sounds right to you. I use 700 Hz. Naturally, it also sets the sidetone frequency of your transmitted signal. As you adjust the control the CW pitch is displayed on the usual popup.

> **CW keyer messages**

There are five CW messages which can be used for DX or Contest operation or just to save you sending the same information over and over. They are great for sending CQ on a quiet band. To use them, select CW mode and then <FUNC> <REC/PLAY>.

If you have the external FH-2 keypad plugged into the REM jack, you can send the five CW messages by pressing buttons 1 to 5. As always, BK-IN must be on for the CW signal to be transmitted. Provided at least one CW Memory slot is set to MESSAGE, you can press the MEM button to record a message from the paddle into that slot. This is useful for short-term storage, for example, to hold the other station's callsign.

If you have inserted the # character after the signal report into any of the TEXT messages. The 'contest' number will automatically increment each time the message is sent. If you need to send the same number again, you can decrement the number with the DEC Soft Key in the REC/PLAY window, or by pressing the DEC button on the FH-2 keypad. The message keyer setup instructions are on page 21.

> **CW decoder screen**

I was very excited about the CW decoder because I am hopeless at Morse code. It works reasonably well although for best decoding results you should set the speed of the CW keyer to something close to the speed of the code that is being received. To turn it on use <FUNC> <DECODE>.

I have a few grumbles which are common to all of the decoder screens.

1. There are five unused Soft Key buttons that should be used for the five memory keyer messages but are not.
2. There is no 'Clear Screen' function.
3. You cannot use an external keyboard to type messages.
4. You cannot see the spectrum display when the decode screen is active, or switch between the two screens easily. If you close the decode window to see the spectrum display, you have to go back through the FUNC menu to get it back again. You can display a screen with the spectrum display for the sub receiver and the decode screen for the main receiver at the same time. Which is of no use at all.
5. The message sending popup window blocks the received text. Also, it disappears if you touch almost any other control.

TIP: You can shift the decoder to the sub-receiver using <FUNC> <OPERATION SETTING> <GENERAL> <SUB> (the default is Main).

OPERATING SPLIT

Working in the 'Split' mode is a very common requirement if you are trying to work a rare DX station or DXpedition when they have a 'pileup' of stations calling them.

To work 'Split' you usually set your main-VFO to receive the frequency that the DX station is using and transmit on the frequency indicated by the sub-VFO. On bands above 10 MHz, you transmit USB on a frequency that is a few kHz higher than the DX station. For bands lower than 10 MHz, you transmit LSB on a frequency that is a few kHz lower than the DX station.

For CW the operation is the same, but the split offset is less. Usually, 1-2 kHz rather than 5-10 kHz. If you are the rare DX station or on a or DXpedition you would, of course, reverse the split.

You can operate Split on digital modes, but it is less common. WSJT does it automatically for FT8 operation.

Split operation in the FTDX101 is different to many transceivers where a pre-defined 5 kHz or similar offset is applied. In the FTDX101 using Split means that the transceiver will transmit on the frequency indicated by the sub-VFO display. That might be on a completely different band and mode, so you do have to be careful. Split operation is indicated with an orange LED above the SPLIT button and the red TX LED moves to the sub-VFO side of the VFO knob.

Note you can set the transceiver to operate split without using the SPLIT button. You could use the TX clarifier to create a split of up to 9 kHz from the VFO frequency. Or you could set the sub-VFO frequency without pressing the SPLIT button and select the right

TX button to transmit on that VFO. There is no TXW button, but you can turn on the sub-VFO receiver to hear what is on the proposed transmit frequency.

➢ **Quick split input**

<FUNC> <OPERATION SETTING> <QUICK SPLIT INPUT> (ON or OFF)

If you have QUICK SPLIT INPUT set to on, and you press and hold the SPLIT button, a dialogue box pops up for you to set the positive or negative split offset you want, in kHz. The default is OFF.

➢ **Setting the pre-set split offset**

<FUNC> <OPERATION SETTING> <QUICK SPLIT FREQ> sets the 'quick split' offset. The default offset is +5 kHz, but it is adjustable from -20 kHz to +20 kHz using the MULTI knob. If QUICK SPLIT INPUT is set to off and you press and hold the SPLIT button, the sub-VFO is moved by the amount you selected. If you press and hold the SPLIT button again, a further offset of the same amount is applied. For example, +5 kHz, +5 kHz, +5 kHz.

➢ **SPLIT button rules**

1. Pressing SPLIT only changes the transmitter to the sub-VFO frequency and turns on the orange Split button LED. It does not change the sub-VFO to be on the same band and mode as the main-VFO.

2. If the transmitter was already on the sub-VFO frequency, pressing SPLIT changes the transmitter back to the Main-VFO frequency and turns off the orange Split button LED. It does not change the Main-VFO band and mode.

3. Press and holding the split button when QUICK SPLIT INPUT is set to on, opens a dialogue box for you to set the positive or negative split offset you want, in kHz. It also sets the sub-VFO to the Main-VFO mode and band.

4. Press and holding the split button when QUICK SPLIT INPUT is set to off and Split is off, sets the sub-VFO to the Main-VFO mode and frequency plus or minus the pre-set split offset.

5. Press and holding the split button when QUICK SPLIT INPUT is set to off and Split is already turned on, increments the split offset on the sub-VFO by the pre-set split offset.

6. If both receivers are turned on and you press the SPLIT button to move the transmitter to the sub-VFO frequency, the sub-VFO frequency display turns red to indicate which VFO will be transmitting.

7. When transmitting, the green 'BUSY' squelch indicator above the MHz digit on the VFO becomes a red 'TX' indicator.

➤ A typical scenario

For a typical scenario where I am attempting to contact a DX station using SSB on the 20m band. I recommend the following technique.

- Set QUICK SPLIT INPUT to off and QUICK SPLIT FREQ to +5 kHz using the instructions above. These are the default settings, so you only need to do this if you have changed them previously.
- Turn MONO off (to white). Press DISP for a display with both receiver spectrum scopes, either one above the other or side-by-side.
- If the markers are off, turn them on with <S.MENU> <MARKER>.
- When Split is turned off (no orange LED), press and hold the SPLIT button. This will set the sub-VFO to the same band and mode as the main-VFO with an offset of +5 kHz. *[This will not work if Split is already turned on].*
- Turn on the sub-receiver using the RX button to the top right of the VFO knob. The sub-VFO frequency display will go red. Adjust the squelch and volume of the sub-receiver to your liking.
- Press MAIN/SUB so that the big tuning ring is controlling the sub-receiver's VFO.
- Touch the main-VFO display or press the MAIN button so that the VFO knob is controlling the main-VFO.
- Tune the main-VFO to the frequency that the DX station is transmitting on.
- Temporarily turn down the MAIN AF volume.
- Using the big VFO ring, tune the sub-receiver through the pileup of stations to find a good transmit frequency. Listen to the pileup and try to work out which way the DX station is moving through the pileup or simply use the spectrum and waterfall to find a quiet spot. Make sure that you are transmitting within the span of frequencies the DX station is listening to, for example, "5 kHz up," or "up 5 to 10 kHz."
- Turn down the SUB AF volume and turn up the MAIN AF volume again.
- Now you are ready to make your call.

The technique is similar when you are using the low bands except the pileup will be spread on frequencies below the DX station and you will use LSB. For CW operation the technique is the same except the split will be 1-2 kHz rather than 5-10 kHz.

If you prefer to leave the QUICK SPLIT INPUT setting turned on. When you press and hold the SPLIT button a popup keyboard will appear.

Touch <5> <kHz> to move the sub-VFO 5 kHz high. Or <-> <5> <kHz> to move the Sub 5 kHz VFO low. Personally, I don't think it is worth the bother. I find it easier to have the keyboard function turned off and simply tune the sub receiver frequency using the VFO ring.

> **Operating Split without using the SPLIT mode**

The FTDX101 has two identical receivers, so it doesn't matter which one you use to listen to the DX station and which you use to listen to the pileup. You could turn on both receivers and tune the sub receiver to the DX frequency, then use the main-VFO to tune to the spot in the pileup where you want to transmit. This has exactly the same effect as using the Split button except that the roles of the VFOs are reversed. Use MAIN/SUB to allocate sub-receiver tuning to the big VFO ring.

> **Operating Split with a single VFO**

If you prefer doing it the old-fashioned way, with a single VFO. You can set the radio for a single VFO and simplex operation. MONO set to blue. Then use the TX Clarifier and CLAR to offset the transmitter by anything up to ±9 kHz. With CLAR turned on you can set the transmitter offset using the big VFO ring. Which has a similar effect to using the same control to tune the sub-receiver VFO frequency, but with a more limited range.

SO2R / SO2V OPERATION

You can achieve SO2R (single operator two radio) or more correctly SO2V (single operator two VFO) operation with the FTDX101. Use the main-VFO to work stations on one band and use the sub receiver on the same or a different band to find other stations. You can view dual spectrum and waterfall displays to see the signals on two bands at the same time, or on different parts of the same band. Simply select the right-side TX button when you want to transmit on the sub-VFO frequency and the left-side TX button when you want to transmit on the main-VFO frequency. The main difficulty, apart from listening to two conversations at once, is making sure that you are transmitting on the right band.

The SO2V technique is used by contesters. It can be remarkably effective especially when the bands are quiet and there are few responders to your 'CQ Contest' calls. You can combine SO2V operation with the use of the voice or CW keyer to operate as a 'Run' station on one band while using 'Search and Pounce' on another. Use the message keyer to send out regular 'CQ Contest' calls and tune around with the sub receiver looking for contest stations. When you want to transmit on the sub-VFO band, hit the sub-VFO TX button and transmit. You cannot swap the TX side while the message keyer is transmitting a message and you cannot operate 'split' and SO2V at the same time, because both operating styles use the sub-receiver.

OPERATING CW OR SSB ON TWO BANDS

See SO2V operation above. You can listen to two bands at the same time and just press the relevant TX button when you want to transmit on the alternate band. Set both receivers to the mode of choice, usually CW or SSB.

OPERATING INTERNALLY DECODED CW, RTTY, OR PSK ON TWO BANDS

See SO2V operation above. You can listen to two bands at the same time by turning on both receivers with the two RX buttons. You have to change a menu setting to make the decoder work on the sub-receiver, so for PSK, or RTTY, or CW operation with the internal decoder, it is easier to swap the A and B VFOS using the MAIN◇SUB button directly above the VFO knob when you want to transmit on the alternate band.

OPERATING DIGITAL MODES (FT8 ETC.)

Digital mode operation is easy once you have established the connection between the radio and the computer and set up the com port and audio settings. At the time I am writing this book setting the options for the new PRESET mode is buggy and likely to cause the transceiver to freeze. I am sure that it will be fixed soon.

FT8 has swept the world as it has allowed modest-sized stations the opportunity to work DX when band conditions have been very marginal due to the sunspot cycle. As the new cycle ramps up over the next few years, I hope we will see many people return to 'conversation' modes such as PSK, Domino, and RTTY.

I like to set up all my digital mode programs so that they all use the same settings on the radio. It makes changing between them so much easier. I use MixW for digital modes, WSJT for FT8, and MRP40 for CW (receive only). I can't go into the operation of these programs. That would be a whole other book! But I have included setup instructions for some of the popular programs.

➢ **Operating with external digital mode software**

Turn on your digital modes program. I recommend using the main-VFO for digital mode operation.

Press and hold MODE and select the mode. Usually DATA-U.

If you have set up the Preset mode, press and hold MODE and select Preset. The icon must be blue (not grey).

Select the frequency and band of operation. WSJT and some other programs have band dropdown menus or buttons which will automatically place the transceiver on the correct frequency. WSJT will also use the sub-VFO to transmit using a higher audio frequency. This stops harmonics of the audio signal creating false FT8 signals across the band.

Make sure that the audio level into the digital modes program is sufficient, but not overloading the program. If necessary, use the program's soundcard settings or the Windows sound settings to set a suitable level. There is no audio output level on the transceiver.

Make sure that the transmit audio level into the radio is sufficient to modulate the radio to full power (or a lower power if you prefer). The ALC meter should be very zero or very low to avoid compression of the digital mode signal. If necessary, use the program's soundcard settings or the Windows sound settings to set a suitable level. You can adjust the audio level on the transceiver, but I prefer to leave it set the same so that all digital mode software is set up using the program settings instead of changing settings on the radio.

The digital mode program should control the PTT. Enjoy operating on the digital mode.

➢ **Operating with the internal decoders**

You don't need any connection with the PC to use the three modes that have internal decoders and message macro keyers.

I recommend using the main-VFO for digital mode operation. There are no levels to adjust or worry about. Press and hold MODE and select the mode PSK, RTTY-L, or CW.

You should not use the Preset for the internal digital modes. Press and hold MODE to check that the Preset Soft Key icon is grey (not blue). If necessary, touch PRESET to turn it off.

Select the frequency and band of operation, then tune into a signal, to check that the decoder is working. On CW you can use the tuning indicator on the DSP filter function display, or ZIN, and the APF filter to clean up the CW signal. On RTTY use the mark and space tuning indicator lines on the DSP filter function display to tune into an RTTY signal. On PSK use the 'C' marker.

Now open the decoder screen using <FUNC> <DECODE>. It will overlay the spectrum and waterfall display. You can adjust the DEC LVL to minimize rubbish decodes. To close the decoder, touch the DEC OFF Soft Key.

When you are ready to transmit one of the five messages. Press <FUNC> <REC/PLAY>. Unfortunately, this will sit over the top of the Decode screen so you can't see what you are responding to. If you are planning on operating this way regularly you should definitely invest in an FH-2 keyboard or build your own using the schematic in this book. You can do a quick edit to add a callsign or other information by touching MEM then the message number. Touching the screen outside of the REC/PLAY popup will close the keyer screen. You have to go back to the FUNC menu to get it back again.

FM REPEATER OPERATION

Repeater operation is best achieved by tuning in a repeater, setting all the relevant offset and tone squelch requirements, then saving the channel into a memory slot. If you set a repeater offset or tone frequency in VFO mode, it will stay set, over the whole band, until you change it back. There is no auto-repeater mode on this radio.

Use the MODE button to select FM mode and tune the main-VFO to the repeater output frequency.

Press <FUNC> <RPT> then turn the MULTI knob to set a plus or a minus repeater offset. You have to be quick! The selection popup only stays in the RPT mode for a few seconds. There is a choice of, SIMP (simplex no offset), + (plus offset), and – (minus offset). With a plus offset the transmitter will transmit higher than the receive frequency. With a minus offset the transmitter will transmit lower than the receive frequency. The setting is stored separately for each band.

> Setting the FM repeater offset

You can set the amount of offset for the 6m and 10m bands in the main menu, but not whether it is a plus or minus offset. *Why not?*

The repeater offset for the 28 MHz (10m band) can be set in 10 kHz increments between 0 and 1000 kHz. It would have been handy if the offset could be set from -1000 kHz to +1000 kHz. The default offset is 100 kHz which will suit the vast majority of repeaters.

<FUNC> <RADIO SETTING> <MODE FM> <RPT SHIFT (28MHz)> (Def 100 kHz)

The FM offset for the 50 MHz (6m band) can be set in 10 kHz increments between 0 and 4000 kHz. It would have been handy if the offset could be set from -4000 kHz to +4000 kHz. The default is 1000 kHz which will suit the vast majority of repeaters. All New Zealand 6m repeaters use a -1 MHz offset, but some other countries use a 500 kHz offset.

<FUNC> <RADIO SETTING> <MODE FM> <RPT SHIFT (50MHz)> (Def 1000 kHz)

> Setting the TONE

Most amateur repeaters use a tone squelch system to prevent the repeater from being held on by noise or interfering signals. Most often they also transmit the same tone on the repeater output frequency so that you can enable tone-controlled squelch of your receiver.

Press <FUNC> <ENC/DEC> to set the CTCSS tone. You have to be quick! The selection popup only stays in the ENC/DEC mode for a few seconds. There is a choice of, OFF, ENC, or TSQ. The setting is stored separately for each band.

- OFF no tone is transmitted.

- ENC a tone is transmitted to open the repeater squelch

- TSQ a tone is transmitted to open the repeater squelch and the same tone must be received to open the squelch on the FTDX101.

The 'sub-audible' CTCSS tones are between 67 Hz and 254.1 Hz. Below the normal 300 Hz – 3 kHz audio range of your receiver. This means that they can be used for signaling, but you won't hear them unless you reduce the L CUT FREQ of the audio low pass filter. The most often used tone is 67 Hz, but your local repeaters may use different tones.

Set the CTCSS tone frequency using <FUNC> <TONE FREQ> and turning the MULTI knob. You have to be quick. The selection popup only stays in the TONE FREQ mode for 4 seconds.

CTCSS Tones (Hz)				
67.0	69.3	71.9	74.4	77.0
79.7	82.5	85.4	88.5	91.5
94.8	97.4	100	103.5	107.2
110.9	114.8	118.8	123.0	127.3
131.8	136.5	141.3	146.2	151.4
156.7	159.8	162.2	165.5	167.9
171.3	173.8	177.3	179.9	183.5
186.2	189.9	192.8	196.5	199.5
203.5	206.5	210.7	218.1	225.7
229.1	233.6	241.8	250.3	254.1

➢ **Operating through a repeater**

OK, so you have set the FM mode, tuned to the repeater output frequency, decided whether you want tone squelch, and set a plus or minus offset.

- With a plus offset your transmitter will transmit higher than the receive frequency because the repeater's input frequency is higher than its output frequency.

- With a minus offset your transmitter will transmit lower than the receive frequency because the repeater's input frequency is lower than its output frequency.

Previously you ensured that the offset frequency is correct for the repeater and set the CTCSS tone frequency.

Now you can transmit and hopefully hear a 'kerchunk' when you click the microphone PTT. (Assuming the repeater has a 'tail.' Not all of them do).

TIP: All these settings are a pain to set up and they stay set, so they are still active when you tune down to a different part of the band. It is highly recommended that once you have verified that the repeater settings are working, that you save the current VFO to a memory channel. Then deactivate the offset back to Simplex and set the tone squelch back to OFF.

➢ **Memory channels and FM**

There is no specific memory storage for FM channels. You can use one of the standard memory channels and name it with the repeater identification.

To store a channel: Tune to the repeater output frequency and set the offset and tone as above. Press V>M to bring up a list of memory channels. Use the MULTI knob or up and down touch screen controls then touch the slot you want to use.

Hold down V>M until you hear a double beep and you will see the frequency pop into the selected memory slot.

Now, before you exit, there are some changes you can make.

- Touch NAME to enter the repeater name. Remember to exit the keyboard with <ENT>.
- Touch MODE to use the MULTI control to change the mode. I don't know why you would want to do that.
- Touch SCAN MEMORY to include the channel in the scan mode or skip it. Use the MULTI control to change the entry.
- Touch DISPLAY NAME to display the stored name instead of the frequency in the VFO frequency display. I rather like this one.
- Save your changes with <WRITE>.

➢ **FM and FM-N modes**

Use the FM mode rather than the FM-N mode for FM operation, unless you are using the radio with a transverter for satellite operation or on frequencies where the narrow band FM-N mode may be required.

The I.F. bandwidth is fixed at 16 kHz on the FM and DATA-FM modes and 9 kHz on the FM-N and D-FM-N modes. The transmitter deviation is ± 5kHz on the FM and DATA-FM modes and ±2.5 kHz on the FM-N and D-FM-N modes.

You cannot adjust the roofing filter or the I.F. bandwidth while the radio is in FM mode. The roofing filters are switched out leaving a 12 kHz RF passband on the spectrum display.

COMPUTER MOUSE OPERATION

Connecting a computer mouse is easy. Just plug a USB mouse into one of the front panel USB ports and you are ready to go. The Logitech M310 and M305 wireless models will work, and probably some other mice. The right mouse button has no function and sadly the mouse wheel does nothing either. It could have been used to scroll the frequency… but no.

So, what can you do with the mouse? These items are not in the manual.

- The left mouse click has the same effect as touching the screen with your finger. You can change the meters, the DSP filter function display, the waterfall to spectrum ratio, the frequencies, and use the Soft Keys.
- If you are showing both spectrum and waterfall displays, you can change the focus to either and change the frequency by clicking on the spectrum display.
- If you press FUNC, you can use the mouse to select any of the menu functions. Annoyingly you can't click the mouse on the Soft Key above the FUNC button or use a right mouse click, to open the FUNC menu, which would have made mouse operation much better.

USB KEYBOARD OPERATION

You can connect a USB keyboard to the radio and use it to fill in the TEXT messages. However, I cannot see much point in plugging an external keyboard when there is an onscreen keyboard provided for this purpose. Unfortunately, you can't use an external keyboard in conjunction with the decoders to send text directly from the keyboard.

TIP: Another odd feature is that if I plug in my Logitech wireless keyboard and mouse combo. The keyboard works, but the mouse doesn't.

SCAN MODE

Selecting <FUNC> <SCAN> starts a scan on the currently selected VFO. Touching the VFO display or the spectrum scope stops the scan.

You can also initiate a scan by press and holding either of the microphone UP or DWN buttons. Press again to stop the scan. You can turn this function off with <FUNC> <OPERATION SETTING> <GENERAL> <MIC SCAN> (ON or OFF).

Starting a scan while in memory channel mode scans through the saved memory channels. Channels can be 'skipped' (excluded from the scan), by changing the setting in the memory channel list. Press M>V and use the up and down arrows, or the MULTI knob, to find the channel, then change the SCAN MEMORY setting from SCAN to SKIP. Touch WRITE to save the changes.

TIP: You can change the scan direction by rotating the VFO tuning knob about a quarter of a turn during a VFO scan. Anti-clockwise tunes down, clockwise tunes up. This is useful if you scan over a signal and want to go back to it.

Set the squelch control so that the receiver audio is just muted. In the SSB and CW modes, the scan rate will slow down if a signal that opens the squelch is received. In all of the other modes, the scan will stop when a signal that opens the squelch is received. You can set what happens when the scan stops on a signal. TIME starts the scan again after 5 seconds. PAUSE halts the scan until the signal drops out and then the radio resumes scanning. <FUNC> <OPERATION SETTING> <GENERAL> <MIC SCAN RESUME> (PAUSE or TIME).

TIP: For SSB and CW, the radio will scan in 10 Hz steps. If the FAST tuning rate has been turned on, the radio will scan in 100 Hz steps. The scanning step size is unaffected by the FINE TUNING button. The step size for FM is 5 kHz for normal tuning and 10 kHz at the FAST tuning rate.

PROGRAMMABLE MEMORY SCAN (PMS)

The radio supports up to nine pre-programmed scan ranges, using nine pairs of 'special-purpose memory pairs' known as L/U pairs.

> **Setting up an L/U pair**

Tune to the lower end of the frequency range that you want to save.

Select V>M to activate the MEMORY CH LIST. Navigate all the way down to the bottom of the 99 memory channels until you reach the M-P1L and M-P1U memory pair. Use the MULI knob. Don't use the up and down arrows. It will take forever because the Yaesu scroll bars don't work.

Touch the M-P1L slot (or whichever L/U slot you want to program). Make sure that you have the right mode selected, then press and hold the V>M button to store the current VFO frequency into the slot. You can name the range if you want to. Then touch WRITE to save the record.

Now tune the VFO up to the high end of the frequency range that you want to save.

Touch the M-PxU slot that corresponds to the L/U pair you have chosen. Press and hold the V>M button to store the current VFO frequency into the slot. Then touch WRITE to save the record. Exit using BACK or FUNC.

> **Starting a PMS scan**

First, you have to select the M-P1L (or another) PMS scan edge. Select M>V to activate the MEMORY CH LIST. Navigate down to the L/U channels using the MULI knob.

Select the L channel then press (don't press and hold) the MULTI knob to set the radio into memory channel mode. Turn the VFO knob clockwise to set the radio to PMS. It is indicated above the orange mode indicator. Then start the scan using <FUNC> <SCAN> or the UP / DWN buttons on the microphone.

TIP: If you want to scan down from the high frequency. Select the U channel and turn the VFO knob anti-clockwise to set the radio to PMS. Then start the scan using <FUNC> <SCAN> or the UP / DWN buttons on the microphone.

The same rules apply as when you are using the normal scan function. You can change the direction of the scan by turning the VFO knob a quarter turn in the direction that you want the scan to go. The scan will slow on SSB and CW and stop on the other modes. See Scan Mode above.

Touch the screen to stop the scan. To escape the PMS mode press V/M to get back to M-P1L, then press V/M again to get back to VFO.

LIMITING THE TUNING RANGE

The PMS mode described above limits the range that you can tune the VFO. For example, if you set a PMS L/U pair for 3.750 MHz to 3.900 MHz, and activate PMS (without starting a scan), you will only be able to tune over that part of the 80m band. This could be handy during a contest as it would stop you from tuning outside of the band of interest. But it only lasts until you change bands. When you return the 80m band the radio is no longer in PMS mode. It reverts to the MT memory tune mode, which is not constrained to the PMS limits.

TIP: If the radio has switched to MT mode, you can press V/M to return to M-P1L. Turn the VFO clockwise and the radio will go back to PMS. Or press V/M again to get to VFO mode.

> **Using PMS to limit the tuning range**

Select M>V to activate the MEMORY CH LIST. Navigate down to the L/U channels using the MULI knob.

Select the M-P1L (or another) PMS scan edge and press (don't press and hold) the MULTI knob to set the radio into memory channel mode. Turn the VFO knob clockwise to set the radio to PMS. It is indicated above the orange mode indicator. Then use the VFO as normal. The radio will not let you tune outside of the PMS range.

To escape the PMS mode press V/M to get back to M-P1L, press V/M to get back to VFO. Or change bands.

USING THE TUNER

The internal tuner is able to match unbalanced loads between 16.7 and 150 ohms. this is equivalent to a SWR less than 3:1. Or 25 to 100 ohms on the 6m band (SWR less than 2:1). The tuning range is typical for most antenna tuners installed inside transceivers and will let you match out to the band edges on 'resonant' antennas such as half-wave dipole or Yagi antennas. The tuner may struggle to tune highly reactive loads. The tuner does not operate on the 70 MHz (4m) band.

You should not use the internal tuner and an external tuner at the same time. If you are using an external antenna tuner, or a linear amplifier followed by an external tuner, they should present the radio with an acceptable termination. Leave the internal antenna tuner turned off.

Pressing TUNE will turn the antenna tuner on. An orange LED on the button indicates that the tuner is active. The tuner will choose settings that were saved after a manual tune operation was performed on a frequency within the 10 kHz window. If no settings have been saved for the current frequency, the tuner will choose settings that were saved after a manual tune operation was performed on a frequency close to the current frequency.

Press and hold TUNE to manually activate a tuning sequence. This will radiate some power into the antenna, so please make sure nobody is using the frequency, before tuning. The tuning sequence takes a few seconds. The radio will always perform a full tune, ignoring any saved information. If you have the SWR meter active you will see the SWR come down as the tuning operation nears completion.

TIP: It is not necessary to do a full tune every time you transmit on a different frequency. The tuner will automatically return to the settings it used the last time you transmitted near the selected frequency. It stores tuning information for each 10 kHz window.

➢ **ATU memories**

After a manual tune operation has been activated, the antenna tuner will store the tuning settings provided that the SWR at that frequency, after tuning, is less than 2:1. On the HF bands. If the resulting SWR is higher than 2:1 but less than 3:1, the tuner will be able to match the load. But the settings will not be stored. If the SWR is greater than 3:1 the tuner will probably not be able to find a match.

If you change to a frequency that is outside of the 10 kHz band segment, the tuner will revert to the settings for the new band segment.

➢ **Rules for the antenna tuner**

1. Check the SWR meter while transmitting. If the SWR is high, you should stop transmitting and start a 'tune' operation by press and holding the TUNE button.

2. Pressing the TUNE button will turn the tuner on. It will set the tuner to capacitance and inductance settings that previously resulted in a match. If the antenna is the same and you use a frequency that the antenna tuner

has tuned before, you should have a well-matched system. Check the SWR meter while transmitting to make sure that the meter hardly moves.

3. The antenna tuner will not automatically initiate a 'tune' operation if the SWR is still high. You have to manually start a 'tune' by press and holding the TUNE button.

➢ **External antenna tuners**

You can select the internal antenna tuner or specify which of the three antenna connectors is connected to an external antenna tuner. <FUNC> <OPERATION SETTING> <GENERAL> (INT, EXT1, EXT2 or EXT3). The EXT settings are only applicable if you are using a Yaesu external antenna tuner. The manual has information about connecting a Yaesu FC-40 remote-mounted external antenna tuner. Note the requirement for a ferrite core isolator on the control cable. See page 112 of the Yaesu Operating Manual. © Yaesu.

USING A LINEAR AMPLIFIER

➢ **Yaesu VL-1000 linear amplifier**

If you are connecting a Yaesu VL-1000 linear amplifier, the PTT switching, automatic band switching, and ALC control are all managed over a CT-178 cable. It connects the 15 pin 'Linear' connector on the FTDX101 to the 'Band-Data 2' and 'ALC 2' connectors on the amplifier. The connections are covered very well in the Yaesu Manual. You should set up the ALC as covered below.

➢ **Other linear amplifiers**

If you are connecting a different linear amplifier, you can use the control signals and band data available on the 15 pin 'Linear' connector, but most people use the easier option using the TX GND and Ext ALC RCA connectors.

TIP: I use a stereo RCA to RCA cable. They are the type used for hooking up your component Hi-Fi system and are available online or at any electronics store.

Stereo RCA cables often have red and white connectors. I use the red lead to connect the 'TX GND' on the radio to the 'PTT' input on the linear amplifier. The white lead goes from 'Ext ALC' on the FTDX101 to 'ALC Out' from the amplifier. You may have to cut a plastic bead to split the connectors apart. (Or use two RCA cables).

TIP: If you have an amplifier and a tuner, the TX GND lead will go to the PTT input on the tuner and a second cable will connect from the PTT output on the tuner to the PTT input on the amplifier.

➤ **Linear amplifier ALC**

The ALC (automatic level control) voltage sent from the linear amplifier to the transceiver automatically reduces the transmitted power so that the linear amplifier is never overdriven, which could cause audio compression or clipping, a protection trip, or possible failure of the linear amplifier. The ALC should be configured so that it is not operating unless the transceiver is accidentally left at full power when driving the amplifier. Don't use it as a method of controlling the amplifier power. It should only be used as a failsafe in the event of a power setting mistake. The FTDX101 ALC line accepts a voltage between 0 and -4 Volts. Check that your amplifier has a compatible output level.

Note: Linear amplifier ALC is completely different from the microphone ALC measured by the ALC meter. Microphone ALC is used to ensure that the modulator is not overloaded by your speech when you are using SSB.

➤ **Adjusting the linear amplifier ALC level**

The common practice of running your transmitter at full power and using the linear amp ALC to ensure that the linear amplifier is not overdriven is a bad idea. The linear amplifier ALC is supposed to be used as a protection method. It is too slow to react to sudden changes in input level or help much with transmitter overshoot.

This test requires you to transmit at the full power of the linear amplifier. If you have a high-power load you can test into that. Otherwise, use your normal antenna and conduct the test at a time or on a frequency that won't interfere with other users.

1. Set the transmitter power well below the level required to drive the linear amplifier to full power. <FUNC> <TX POWER>.

2. Set the radio to a 'full power' mode. FM, or CW with the Keyer turned off.

3. Set the linear amplifier ALC control fully off, (probably fully clockwise). It may be a potentiometer on the rear panel of the amplifier or a software control.

4. Transmit into the linear amplifier and slowly increase the RF power until the linear amplifier reaches full power. Stop transmitting for a while to give the amplifier a rest.

5. While transmitting, change the ALC control until the output power of the transmitter and linear amplifier just begins to reduce. Then back off the control slightly until the linear amp is back to full power and the ALC is not having any effect. At that point, the ALC is not operating during normal transmission but will respond to protect the linear amplifier if you accidentally leave the transmitter at full power.

Operating the radio | 101

> **Yaesu band switching**

Band switching information is available on the 15 pin 'Linear' D-Sub connector. The format is the standard BCD (binary coded decimal) Yaesu format.

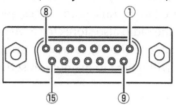

1	+13.5 volts	6	Band Data C	11	TX REQ
2	TX GND (PTT out)	7	Band Data D	12	-
3	Ground	8	TX Inhibit	13	-
4	Band Data A	9	Ground	14	ALC Input
5	Band Data B	10	-	15	Ground

Band	Frequency	DCBA	Hex
160m	1.8 MHz	0001	01
80m	3.5 MHz	0010	02
60m	5 MHz	0011	03
40m	7 MHz	0011	03
30m	10 MHz	0100	04
20m	14 MHz	0101	05
17m	18 MHz	0110	06
15m	21 MHz	0111	07
12m	24.8 MHz	1000	08
10m	28 MHz	1001	09
6m	50 MHz	1010	0A

Memory groups and channels

The FTDX101 has 99 normal memory slots which can be arranged into five groups. There is a 6th group allocated to the nine scan ranges, see PMS on page 96. The U.S. (International) version of the radio also stores ten 5 MHz frequencies which cannot be erased. Memory channel one (M-01) cannot be erased but can be overwritten. If the radio is reset, M-01 will return to the default setting of 7.000 MHz LSB.

➢ **Saving to a memory channel**

Saving a frequency to a memory channel is very easy. Set the mode and tune to the frequency you want to save. For FM channels, tune to the repeater output frequency and set the repeater offset and CTCSS tone. Press V>M to bring up a list of memory channels. Use the MULTI knob or the up and down touch screen controls to find a suitable memory position. Or touch the wanted memory slot. Hold down V>M until you hear a double beep and see the frequency pop into the memory slot.

Before you exit, there are some changes you can make.

- Touch NAME to enter the repeater or channel name. Remember to exit the keyboard with <ENT>.
- Touch MODE and use the MULTI control to change the mode. I don't know why you would want to do that, but you can.
- Touch SCAN MEMORY to include the channel in the scan mode or skip it. Use the MULTI control to change the entry, (SCAN or SKIP).
- Touch DISPLAY NAME to display the stored name instead of the frequency in the VFO frequency display.
- Save your changes with <WRITE> before you exit.

Press the V>M button again to return to the normal display.

➢ **Recalling a saved frequency**

Recalling a saved frequency from a memory channel is easy as well.

Press M>V, or V>M, or press and hold V/M, to bring up the list of memory channels.

Use the MULTI knob or the up and down touch screen controls to find the saved memory channel you want to select. Or touch the wanted memory slot. When the channel you want is highlighted, press (don't press and hold) the MULTI knob.

Pressing V/M returns the transceiver to VFO mode.

Memory channels and groups | 103

> **Stepping through saved memory channels**

Press V/M to select memory channel mode. Then press <FUNC> <MEM CH> and use the MULTI knob to step through the saved memory sots.

> **Turning on memory groups**

Turn saving to memory groups on using <FUNC> <OPERATION SETTING> <GENERAL> <MEM GROUP> (on or off)

To allocate a frequency to a group. You must save the frequency into a memory slot that lies within the group.

> Channels M-01 to M-19 are in Group 1
> Channels M-20 to M-39 are in Group 2
> Channels M-40 to M-59 are in Group 3
> Channels M-60 to M-79 are in Group 4
> Channels M-80 to M-99 are in Group 5
> PMS channels P1 L/U to P9 L/U are in Group 6

You can force the memory mode to only display memory channels within a certain group. But the process is so complicated I doubt many people will bother. It is probably better just to arrange your most used channels near the top of the memory stack.

> Press V/M to change to memory channel mode.
> Press <FUNC> <GROUP> and use the MULTI knob to find the group.
> Press the MULTI knob to select the group.
> Press <FUNC> <MEM CH> and use the MULTI knob to find a channel within the selected group.

> **MT memory tune mode**

If you select a memory channel and then turn the VFO knob, the radio will tune off the saved channel frequency in the same way as when you are using the VFO mode. This is indicated with MT in place of the M-xx channel number above the mode display.

Pressing the V/M button returns the radio to the saved memory channel frequency. Pressing the V/M button again returns the radio to VFO mode.

> **Memory channel scan**

Starting a scan with <FUNC> <SCAN>, or the microphone up/down buttons, while in the memory channel mode, scans through the saved memory channels. Channels can be 'skipped' (excluded from the scan), by changing the 'Scan Memory' setting in the memory channel list. Press M>V and use the up and down arrows, or the MULTI knob, to find the channel, then change the SCAN MEMORY setting from SCAN to SKIP. Touch WRITE to save the changes.

If 'Memory Groups' has been selected only 'non skipped' channels in that group will be scanned. If 'Memory Groups' is turned off, all 'non skipped' channels will be scanned.

➢ **Backup the memory list to the SD card**

Firmware updates or resetting the radio may delete your saved memory channels, so it is a good idea to save a backup onto the SD card.

Make sure that there is an SD card in the SD card slot. Then select,

<FUNC> <EXTENSION SETTING> <SD CARD> <MEM LIST SAVE> <DONE> Select <NEW>. You can change the filename if you want to, but I just touch ENT to select the default date format as the filename.

The radio will format the data file and save it to the SD card. Touch the 'FILE SAVED' icon to return to the previous menu.

TIP: You might as well save a backup of the menu settings at the same time. Touch <MENU SAVE> <DONE> <NEW> <ENT>

Then press FUNC to exit.

➢ **Recalling the memory list from the SD card**

If a radio reset, or a firmware update, wipes out your saved frequencies, you can recover the memory list from your saved backup on the SD card.

Make sure that there is an SD card in the SD card slot. Then select,

<FUNC> <EXTENSION SETTING> <SD CARD> <MEM LIST LOAD> <DONE>

You will be presented with a list of previously saved files. Unless there has been a data corruption problem, you would normally select the newest file, which will be at the bottom of the list. Touch the file name and answer 'OK' to overwrite the memory slots in the radio with the ones stored in the selected file.

Touch the 'FILE LOADED' icon. The radio will re-boot to load the memory channels.

http://clipart-library.com/

Touchscreen display functions

This section deals with the interactive items that are displayed on the touchscreen. It describes what happens when you touch the meter displays, the frequency readout, and the spectrum scope. The next section covers the Soft Key controls.

METERS

Touch either of the meters to select the transmitter meter options. The left meter can display; PO (power out), COMP (speech compression), or TEMP (temperature measured at the power amplifier transistors. The right meter can display; SWR (antenna standing wave ratio), ALC (automatic level control of the modulating audio), V_{DD} (power amp FET drain voltage), or I_D (power amp FET drain current).

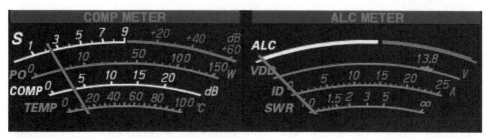

Figure 27: Meter display in MONO single receiver mode

Both meters are displayed in the MONO single receiver mode. (Meter visibility depends on the position of the DSP filter function display).

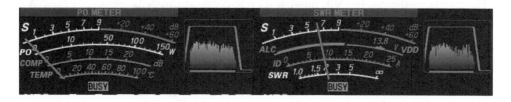

Figure 26: Meter display in dual receiver mode

In the dual receiver mode, the left meter indicates the main receiver S meter, and the right meter indicates the sub receiver S meter. (Meter visibility depends on the DSP filter function display). When you transmit, the left meter usually displays the RF power output and the right meter displays SWR, ALC, or whatever you set.

> **Hidden meters**

The DSP filter function display can hide the meter displays. Touch the filter function spectrum display to restore the meter display.

➤ EXPAND mode

When the expanded spectrum scope is enabled using the EXPAND Soft Key, the meters are compressed to two scales. Showing only the S meter and the chosen transmitter metering instead of all the meter scales. The right meter does not show the sub receiver S meter in the MONO single receiver mode.

Figure 28: Metering when the spectrum scope is expanded

➤ External meters

You can connect a pair of external meters to the 3.5 mm METER connector on the rear panel of the radio. The output ranges from 0 to 3 volts. You can use a series current limiting resistor if you want to connect a mA meter.

DSP FILTER FUNCTION DISPLAYS

The DSP filter function displays show the received spectrum within the receiver passband. The red line outlines the shape of the filter response and the bandwidth of the currently selected R.FIL roofing filter is indicated by the blue line at the bottom of the spectrum display.

Touch and hold the filter function display to toggle the spectrum display off or on.

The display indicates the SSB passband on SSB, the audio peak frequency for CW, the center frequency for PSK, or the Mark and Space frequencies for RTTY. It also indicates the effect of any I.F. Shift and width adjustment, and active contour or notch filters.

TIP: *When you are using the MONO display with a 3 kHz or 12 kHz roofing filter, part of the received passband is hidden by the second meter. Touch the DSP filter function display to see the full receiver passband.*

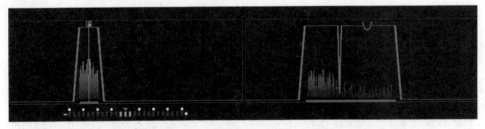

Figure 29: **CW and SSB (with Notch and Contour) Filter Function Displays**

THE FREQUENCY DISPLAY

The VFO frequency is displayed in large white letters. The current mode is displayed to the left with a white indicator above. This indicator reads VFO in VFO mode, M-xx in memory channel, or MT in memory tune mode. In the PMS (programmable memory scan) mode, it indicates the PMS channel number. As soon as you move the VFO tuning knob this changes to PMS. This limits VFO tuning to the limits of the PMS memory range and a scan will only cover the same range.

If the radio is in memory channel mode and 'Display Name' has been set to display a name, it will be displayed instead of the frequency digits. Turn the MULTI knob to step through the saved memory channels.

The frequency display has several touchscreen functions.

> **Hz digits**

Touch the three Hz digits to enter a frequency using the onscreen keyboard. Touch ENT to load the frequency into the VFO.

TIP: If the entered frequency is above 10 MHz there is no need to touch the decimal point. In fact, the decimal point will not respond if a double-digit number has already been entered.

> **kHz digits**

Touch the three kHz digits to tune the VFO in 1 kHz steps. The display will drop back to the normal tuning rate a few seconds after you stop tuning the VFO.

> **MHz digits**

Touch the MHz digit(s) to tune the VFO in 1 MHz steps. The display will drop back to the normal tuning rate a few seconds after you stop tuning the VFO.

> **Tuning rates**

As you turn the VFO knob the frequency normally changes in 10 Hz steps. Press FAST to increase the tuning rate to 100 Hz per step, or FINE TUNING to change to 1 Hz per step. The normal VFO rate can be changed separately to 5 Hz per step for SSB/CW or RTTY/Data.

<FUNC> <OPERATION SETTING> <TUNING> <SSB/CW DIAL STEP>

<FUNC> <OPERATION SETTING> <TUNING> <RTTY/PSK DIAL STEP>

The normal tuning step for FM is 100 Hz, FAST is 1 kHz, and FINE TUNING is in 10 Hz steps.

If you apply FINE TUNING and FAST at the same time, the FINE TUNING button overrides the FAST button.

➤ STEP DIAL rates

If you have allocated either the MULTI knob or the big VFO ring CS function to STEP DIAL, the FAST button increases the step size from 2.5 kHz to 25 kHz. FINE TUNING has no effect on the STEP DIAL function.

<FUNC> <OPERATION SETTING> <TUNING> <CH STEP> changes the step dial rate for all modes except AM and FM, (1, 2.5, or 5 kHz)

<FUNC> <OPERATION SETTING> <TUNING> <AM CH STEP> changes the step dial rate for AM, (2.5, 5, 9, 10, 12.5, or 25 kHz)

<FUNC> <OPERATION SETTING> <TUNING> <FM CH STEP> changes the step dial rate for FM, (5, 6.25, 10, 12.5, 20, or 25 kHz)

➤ Changing the VFO focus

If you have both VFO displays operating, you can change the focus to the other VFO by touching the frequency display. This is the same as pressing the MAIN and SUB buttons. Note that this does not turn on the sub-receiver. You have to press the appropriate RX button to do that.

➤ FAST (press and hold)

Press and holding FAST resets the three Hz digits to 000. If the scope is set to FIX mode, press and holding FAST resets the VFO to the beginning of the FIX spectrum range.

➤ Optical encoder rate VFO

You can also change the rate at which the VFO knob works by changing the number of steps per revolution. If you select a lower number of steps per revolution, the VFO step size remains the same, but you have to rotate the knob more degrees to make the numbers change. If you select a higher number of steps per revolution, the VFO step size remains the same, but you don't have to rotate the knob as much to make the numbers change. It effectively changes the sensitivity of the control.

<FUNC> <OPERATION SETTING> <TUNING> <MAIN STEPS PER REV> (250, 500, or 1000 steps per revolution).

➤ Optical encoder rate MPVD "big VFO ring"

To change the encoder rate for the MPVD (Multi-Purpose VFO Outer Dial), otherwise known as the "big VFO ring," select,

<FUNC> <OPERATION SETTING> <TUNING> <MPVD STEPS PER REV> (250 or 500 steps per revolution).

SPECTRUM AND WATERFALL DISPLAY

Touching the screen within the spectrum scope area moves the VFO to the frequency you touch. It provides a fast way to navigate to a wanted signal or frequency. You will still have to fine-tune using the VFO knob.

Touching the waterfall display toggles through three options that set the ratio of the waterfall area to spectrum scope area. Roughly 2/3 spectrum and 1/3 waterfall, or 50/50, or 1/3 spectrum and 2/3 waterfall.

If you have dual spectrum displays operating, touching either the 2D spectrum scope or the 2DSS spectrum moves the relevant VFO frequency, but touching the 2D waterfall has no effect. The sub-receiver spectrum and waterfall display are always on the top, or on the right, of the main-receiver spectrum and waterfall display. Each receiver has its own set of ANT, ATT, IPO, R.FIL, and AGC Soft Keys. The lower set of Soft Keys affects whichever VFO has focus, (larger VFO digits). Note that changing the focus does not turn on the sub-receiver. You have to press the appropriate RX button to do that.

You can use a computer mouse to change the same things as touching the screen with your finger or a stylus.

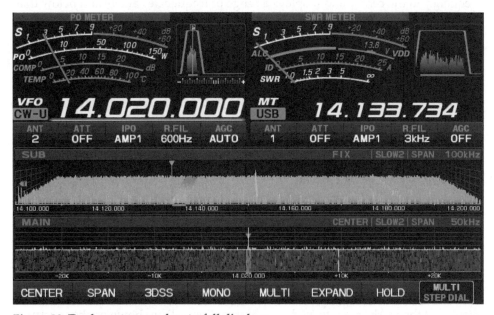

Figure 30: **Dual spectrum and waterfall displays**

Display Soft Keys

This chapter discusses the operation of the Soft Keys. What they do and how to use them. There are seven Soft Keys at the bottom of the display, just above the eight buttons. The eighth position is not a Soft Key it is just an indicator. It shows the setting currently allocated to the MULTI knob.

NOTE: This is a real missed opportunity. The eighth position could have displayed the setting currently allocated to the MULTI knob and also acted as a Soft Key duplication of the FUNC button. That way you would be able to use the mouse to select FUNC which would be super helpful if you like to use a mouse to operate the radio.

Near the middle of the screen, just below the VFO frequency display, there is another row of Soft Keys. Five for each displayed VFO.

THE LOWER SOFT KEYS

The lower Soft Keys are used to set the various display functions. The first Soft Key is used to control the spectrum scope and waterfall display. The **CENTER** option makes the display act as a band-scope like the displays fitted to superheterodyne architecture DSP radios. A band-scope places the receiver VFO frequency in the center and displays a range of frequencies above and below that frequency.

NOTE: The narrowband receiver bandwidth is usually highlighted on the waterfall in a different shade or Color. If you are using the normal 'carrier point' display, it is displayed on the right of center for upper sideband, or on the left of center for lower sideband. The displayed width depends on the selected roofing filter and the span of the display. If you selected 'Filter,' the waterfall display will be across the centerline. If markers are enabled a marker will indicate the carrier point. To change the scope center point, select <FUNC> <DISPLAY SETTING> <SCOPE CTR> and select CAR POINT or FILTER. To toggle markers on or off, select <S. MENU> <MARKER>. [The FILTER option is horrible – don't use it].

The **CURSOR** option behaves more like the spectrum and waterfall on an SDR receiver. Remember that the spectrum display is created by the two direct sampling SDRs. The point where the transceiver is receiving can be anywhere across the display. The receiver passband is indicated on the waterfall in a different shade or Color and a brown bar is shown at the bottom of the waterfall display. If the markers are turned on, a marker line will be shown on the spectrum display at the carrier point frequency. Use <S. MENU> <MARKER>. The green marker indicates the receiver frequency, and the red marker indicates the transmitter frequency. In simplex mode, they overlay each other with the red marker on the top. It is easier to see both markers if you select Split operation. I strongly recommend leaving the markers turned on all the time.

If you tune above or below the range of frequencies in the currently displayed span, the markers will sit at the edge of the display and the entire spectrum and waterfall or 3DSS display will scroll as you tune. Touching the spectrum or waterfall will move the VFO to a frequency close to the position you touched.

The **FIX** display works in a similar way to the CURSOR display, except the behavior when you tune off the edge is different. When you touch FIX there is a good chance that the receiver passband will not be shown on the display at all. Touching the spectrum or waterfall will move the VFO so that it is within the fixed display band. FIX is the only Soft Key that has a touch and hold function. Touch and hold opens up a popup window that allows you to set the frequency that will be at the extreme left side of the spectrum display. Enter the frequency and then touch ENT to save your data. The frequency at the right side of the display is automatically adjusted according to the scope bandwidth set by the SPAN setting.

Unlike the CURSOR display which scrolls if you tune below or above the current scan. The FIX display stays, well… "fixed." If you have markers turned on, small green and red indicators at the right or left indicate which direction the VFO frequency is, compared to the displayed FIX span. Note that the shaded area on the waterfall or 2DSS display that indicates the bandwidth of the signal gets stranded at the edge of the screen and no longer indicates the receiver passband. Touch the spectrum scope to move the VFO frequency back into the fixed panadapter bandwidth.

> **The problem with the FIX display**

The implementation of the FIX display is a major disappointment. It does not work the way I think that it should. I hope that Yaesu will look at the way this mode is implemented on virtually every other SDR receiver or transceiver and issue a firmware update. As it stands, it is hopeless.

On the FTDX101 you can only set a start frequency for the left side of the spectrum display. The frequency at the right side of the display is set according to the currently displayed Span setting. At a minimum, you should be able to set both the high and low, frequency limits for the fixed scope on each band. As a comparison, recent Icom transceivers allow you to set four fixed display regions for each of twelve frequency ranges. For example, on the 20m band, I have one set up to show the SSB segment from 14.100 to 14.350 MHz, the second shows the CW band segment from 14.000 to 14.060 MHz, and the third displays the digital mode part of the band from 14.060 to 14.100 MHz. The only constraint on the Icom radio is you can't have a band segment that is wider than the maximum display bandwidth of 1 MHz.

THE UPPER SOFT KEYS

The radio displays five Soft Keys near the middle of the screen for each displayed VFO. Five are visible on the MONO display, and on the dual VFO displays, all ten are visible.

ANT	ATT	IPO	R.FIL	AGC
2	OFF	AMP1	3kHz	AUTO

ANT: Selects the antenna port on the back of the radio. There is no menu item for saving antenna settings per band, but the radio will switch back to the last used antenna when you switch bands, so that works fine.

ANT1 and ANT2 are straightforward. They select whether the radio will receive and transmit through the ANT1 or the ANT2 connector.

The function of the ANT3 connector is set using <FUNC> <OPERATION SETTING> <GENERAL> <ANT3 SELECT>. There are four options.

- TRX makes ANT3 the same as the other two antenna ports. The radio will transmit into and receive from the antenna connected to the ANT3 connector. If selected, the ANT Soft key will display 3.
- With the R3-T1 option, the transceiver will transmit into the antenna connected to ANT1 and receive from the antenna connected to ANT3. If selected, the ANT Soft key will display R/T1. *Why not R3/T1?*
- With the R3-T2 option, the transceiver will transmit into the antenna connected to ANT2 and receive from the antenna connected to ANT3. If selected, the ANT Soft key will display R/T2. *Why not R3/T2?*
- With the RX-ANT option, transmit operation is disabled and the transceiver will receive from the antenna connected to ANT3. If selected, the ANT Soft key will display RANT.

ATT: Enables the 6 dB and/or 12 dB front-end attenuators. You can choose OFF, 6 dB, 12 dB, or 18 dB.

Tip: hardly anyone uses the ATT control on their HF radio, but it can really improve the signal to noise ratio, especially on the noisy 40m, 80m, and 160m bands. If signals are strong, adding attenuation will make the signal better by cutting out a lot of the background noise. Try it. You will like it!

You can turn the attenuators on and still have AMP1 or AMP2 enabled, but it is preferable to turn the preamplifiers off before selecting an attenuator. Yaesu recommends selecting the IPO mode using the IPO Soft Key and if the problem persists, adding in some attenuation with the ATT Soft Key. You can also try the noise blanker (NB) and digital noise reduction (DNR).

TIP: On my radio, ATT 6dB reduces the signal by 6 dB compared to the IPO setting with an S9 input signal on SSB or CW.

ATT 12 dB reduces the signal by 13 dB and ATT 18 dB reduces the signal by 19 dB. The S meter decreases by two S points per 6 dB step. Activating both AMP2 and the 18 dB attenuator results in close to the same S meter reading as using IPO and no attenuation.

IPO: The IPO Soft Key cycles through three settings. AMP1 (a 10 dB preamplifier), AMP2 (a 20 dB preamplifier), or IPO which turns off both preamplifiers. The 'normal' setting is to leave AMP1 turned on. Normally AMP2 would only be used on the quieter 10m and 6m bands. You might consider using the IPO option for the noisier 40m, 60m, 80m, and 160m bands. The IPO setting is remembered when you change bands or switch the radio off.

According to the Yaesu manual, *"The IPO (Intercept Point Optimization) function maximizes the dynamic range and enhances the close multi-signal and intermodulation characteristics of the receiver."* This sounds mighty impressive, and I expect, like me, you are expecting some "high-tech wizardry." Sadly, all the IPO Soft Key does is turn off both preamplifiers and add around 7 dB of attenuation to the signal. The claim about improving the dynamic range and intermodulation performance is undoubtedly true, but not as exciting as I had expected. Reducing the input level with an attenuator typically increases the dynamic range and signal to noise ratio and turning off the preamplifiers improves the IMD (intermodulation distortion) performance.

If the received noise level is high, Yaesu recommends selecting the IPO mode and if the problem persists, adding in some attenuation with the ATT Soft Key. You can also try the noise blanker (NB) and digital noise reduction (DNR).

TIP: On my radio, AMP1 adds 9.1 dB of gain above the IPO setting with an S9 input signal on SSB or CW. AMP2 adds 17.9 dB of gain above the IPO setting. The S meter increases by 10 dB and 18 dB respectively. Activating both AMP2 and the 18 dB attenuator results in close to the same S meter reading as using IPO with no attenuation.

R.FIL: The receiver's selectivity and 3rd order IMD dynamic range is greatly improved by the sharp narrowband roofing filters working at the 9 MHz Intermediate Frequency. They can be selected with the R.FIL Soft Key or the R.FIL menu command. Generally, you will use the 600 Hz filter for CW and the 3 kHz filter for SSB. The setting will change to the last used bandwidth if you change between CW and SSB.

The FTDX101D has 600 Hz, 3 kHz, and 12 kHz roofing filters fitted as standard. 300 Hz and 1.2 kHz filters can be added but only as a factory-installed (not user installed) option. The FTDX101MP has the same three filters installed, plus a 300 Hz filter on the main receiver only.

The 1.2 kHz and second 300 Hz filter can be added as factory-installed options. There are no DSP filter options although you can use the IF width and shift controls.

AGC: The AGC can be set to OFF (not recommended), AUTO (default), FAST, MID, or SLOW.

You use the MULTI knob to adjust the 'Fast,' 'Mid,' or 'Slow' delay individually for SSB, AM, FM, PSK & DATA, RTTY, and CW modes.

The AGC delay can be set for each of the manual AGC settings in the menu.

<FUNC> <RADIO SETTING> <AGC FAST DELAY>

<FUNC> <RADIO SETTING> <AGC MID DELAY>

<FUNC> <RADIO SETTING> <AGC SLOW DELAY>

The CW setting is in the <FUNC> <CW SETTING> <MODE CW> menu.

Upper Soft Keys				
Setting	Options	FTDX101 default	ZL3DW setting	My Setting
ANT	1, 2, 3 R/T1 R/T2 RANT	ANT1	Depends on the band	
ATT	OFF 6 dB 12 dB 18 dB	OFF	OFF	
IPO	IPO AMP1 AMP2	AMP1	AMP1	
R.FIL	300 Hz (MP) 600 Hz 3 kHz 12 kHz	3 kHz for SSB 600 Hz for CW	3 kHz for SSB 600 Hz for CW	
AGC	OFF AUTO FAST MID SLOW	AUTO	AUTO	

Function menu

All of the menu options and radio settings are accessed by pressing the FUNC button. The various settings are arranged in six groups indicated by Colored lines under the Soft Keys.

GROUP 1: ORANGE: SPECTRUM AND WATERFALL

The S.MENU button brings up the same five items. My settings are in brackets.

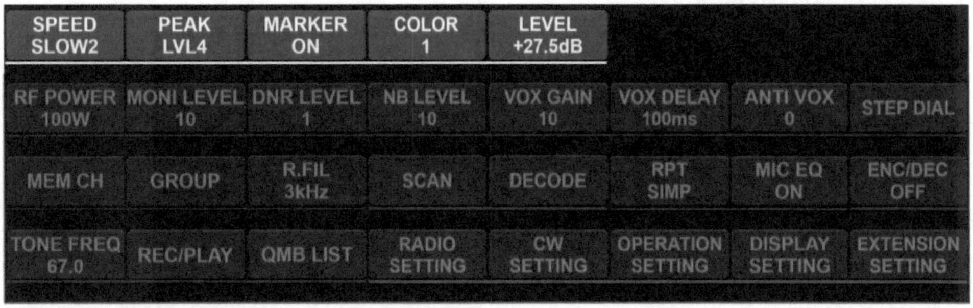

SPEED SLOW2	PEAK LVL4	MARKER ON	COLOR 1	LEVEL +27.5dB				
RF POWER 100W	MONI LEVEL 10	DNR LEVEL 1	NB LEVEL 10	VOX GAIN 10	VOX DELAY 100ms	ANTI VOX 0	STEP DIAL	
MEM CH	GROUP	R.FIL 3kHz	SCAN	DECODE	RPT SIMP	MIC EQ ON	ENC/DEC OFF	
TONE FREQ 67.0	REC/PLAY	QMB LIST	RADIO SETTING	CW SETTING	OPERATION SETTING	DISPLAY SETTING	EXTENSION SETTING	

Figure 31: **The FUNC menu – orange group**

Item	Function	Behavior
Speed	Sets the waterfall speed in 5 stages. (SLOW 2).	The screen disappears but leaves SPEED associated with the MULTI control
Peak	Sets the sensitivity of the waterfall display. Affected by the setting of the spectrum LEVEL. (LVL 3).	The screen disappears but leaves PEAK associated with the MULTI control
Marker	Turns on or off the marker that indicates the receive (green) and transmit (red) VFO frequencies on the spectrum and waterfall display. (ON).	The screen disappears but toggles the marker on or off. I recommend leaving the markers turned on unless you only use the CENTER screen display.
Color	Allows you to choose the waterfall Colors. There is a choice of eleven Colors for the 3DSS display or main waterfall and seven Colors for the area of the waterfall within the selected receiver	The screen disappears but leaves COLOR associated with the MULTI control. Turning MULTI brings up the Color selecting dialogue box. You have to act fast! It only stays visible for a few seconds. Touch and hold M1, M2, or M3 to save the current Color settings.

	bandwidth. Some of the Color selections are graduated according to signal strength.	
Level	Sets the spectrum scope level. The spectrum level changes if you change bands and also depending on how noisy or active each band is.	The screen disappears but leaves LEVEL associated with the MULTI control. If after adjusting the spectrum level you are not happy with the waterfall brightness, adjust PEAK.

TIP: *There is no dedicated control for setting the spectrum level, which is a shame because you adjust it a lot. If you select LEVEL via the FUNC or S.MENU button, the LEVEL function will stay allocated to the MULTI knob until you make another selection. This makes it handy for adjusting the spectrum level. Or you can allocate the spectrum level adjustment to the large ring around the VFO knob by press and holding the CS button and selecting LEVEL. I find this the best use of the VFO ring.*

NOTE: *The 3DSS display shares the same Color selection as the waterfall. The LEVEL and PEAK functions work on the 3DSS display as well. The display seems to auto level, so I was not able to make it peak anywhere near the top of the five-grid display. It can display a weak signal of -109.5 dBm in the SSB bandwidth, the same as the normal spectrum display. The standard spectrum display is nearly always white and there is no averaging function. It is always filled and cannot be set to a line or peak-hold display. There is no gain control for the spectrum display, but it appears to be set to 5 dB per division. On my radio, it displays signals in a 50 dB range between -109.5 dBm to -60.5 dBm.*

GROUP 2: PURPLE: RADIO FUNCTIONS

This group contains some adjustments that would have been front panel controls on earlier radios and a few that may have been buried in the menu structure.

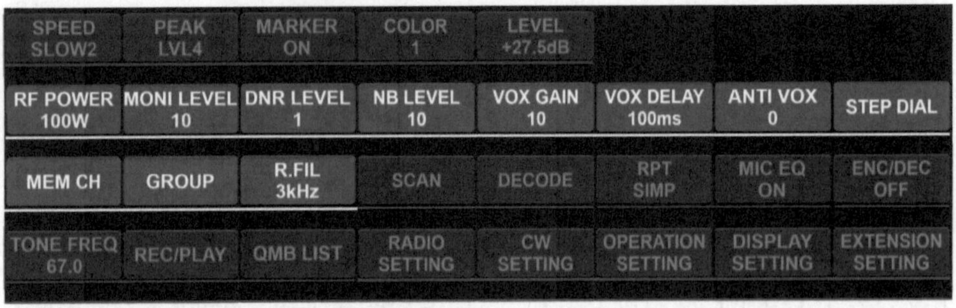

SPEED SLOW2	PEAK LVL4	MARKER ON	COLOR 1	LEVEL +27.5dB				
RF POWER 100W	MONI LEVEL 10	DNR LEVEL 1	NB LEVEL 10	VOX GAIN 10	VOX DELAY 100ms	ANTI VOX 0	STEP DIAL	
MEM CH	GROUP	R.FIL 3kHz	SCAN	DECODE	RPT SIMP	MIC EQ ON	ENC/DEC OFF	
TONE FREQ 67.0	REC/PLAY	QMB LIST	RADIO SETTING	CW SETTING	OPERATION SETTING	DISPLAY SETTING	EXTENSION SETTING	

Figure 32: **The FUNC menu – purple group**

Function menu | 117

All of the functions in this group can be allocated to the large ring around the VFO knob by press and holding the CS button.

Item	Function	Behavior
RF Power	Sets the transmitter power between 5W and 100W for the FTDX101D (25W AM) and 5W to 200W for the FTDX101MP (50W AM). The manual says 50W for the 70 MHz band, but it is deactivated on my radio as we don't have the 4m band in ZL. (100W).	The screen disappears but leaves RF Power associated with the MULTI control.
MONI LEVEL	Sets the monitor level that you hear in the speaker or headphones while transmitting if the MONI button has been selected. (MONI 10).	The screen disappears but leaves MONI LEVEL associated with the MULTI control
DNR LEVEL	Allows you to set the amount of digital noise reduction. It is not really a level control. You can choose one of fifteen different DSP noise-reduction algorithms. (DNR LEVEL 4).	The screen disappears but leaves DNR LEVEL associated with the MULTI control. Adjusting the level automatically enables the DNR button.
NB LEVEL	Allows you to set the noise blanker threshold. (NB LEVEL 4).	The screen disappears but leaves NB LEVEL associated with the MULTI control. Adjusting the level automatically enables the NB button.
VOX GAIN	Allows you to set the VOX gain using the MULTI knob. See more about VOX on page s 79.	The screen disappears but leaves VOX GAIN associated with the MULTI control.
VOX DELAY	Allows you to set the VOX delay using the MULTI knob.	The screen disappears but leaves VOX DELAY associated with the MULTI control.
ANTI VOX	Allows you to set the anti-VOX protection using the MULTI knob.	The screen disappears but leaves ANTI-VOX associated with the MULTI control.
STEP DIAL	Allows you to move the VFO in large increments	The screen disappears but leaves STEP DIAL associated with the

	using the MULTI knob. Default 2.5 kHz steps except for FM and AM which has 5 kHz steps. (1 kHz for SSB).	MULTI control. The FAST button causes the STEP size to be 10 times normal. The STEP Size can be changed see note 1 below this table.
MEM CH	Lets you select one of the 99 memory channels or one of the 60m band 5 MHz channels. See memory channel operation on page 103.	The screen disappears but leaves MEM CH associated with the MULTI control. Note that the control will not change to the selected memory channel unless 'memory mode' is selected using the V/M button. See note 2 below this table.
GROUP	In memory mode, you can use the MULTI knob to select a memory group. It gets you to a group of memory channels fast. See memory channel operation on page 103.	The screen disappears but leaves GROUP associated with the MULTI control. This control does not work in VFO mode. The default setup only has one group, for the 5 MHz channels. But you can add four more. The 6th group is allocated to PL scan edges.
R.FIL	Sets the receiver roofing filter. This can be achieved more easily by touching the R.FIL Soft Key on the bar that separates the spectrum display from the VFO display.	The screen disappears but leaves R.FIL associated with the MULTI control.

NOTE 1: You can change the step size for AM, FM, and all other modes.

I changed the setting for SSB because stations are usually spaced by multiples of exactly 1 kHz. <FUNC> <OPERATION SETTING> <CH STEP> <1 kHz>

I left the AM setting at the default.
<FUNC> <OPERATION SETTING> <AM CH STEP> <5 kHz>

I changed the setting for FM because it is a standard channel spacing.
<FUNC> <OPERATION SETTING> <FM CH STEP> <12.5 kHz>

NOTE 2: Press and hold the V/M button to get a full list of the memory channels. Use the MULTI control to choose a stored frequency then push MULTI to select it. This action will place the transceiver in 'memory mode.' Press V/M if you want to return to VFO mode.

TIP: If you see MT above the mode indicator it indicates that the VFO is in 'memory tune' mode. This occurs if you selected a memory channel and then tuned off the frequency. Press V/M to get back to the selected memory channel frequency. Then press V/M again if you want to get back to VFO mode. Note that you cannot tune off the 60m band 5 MHz preset memory channels.

GROUP 3: GREEN: SCAN AND DECODE

This group contains a Soft Key to start scanning and one that turns on the CW, RTTY, or PSK decoder screen.

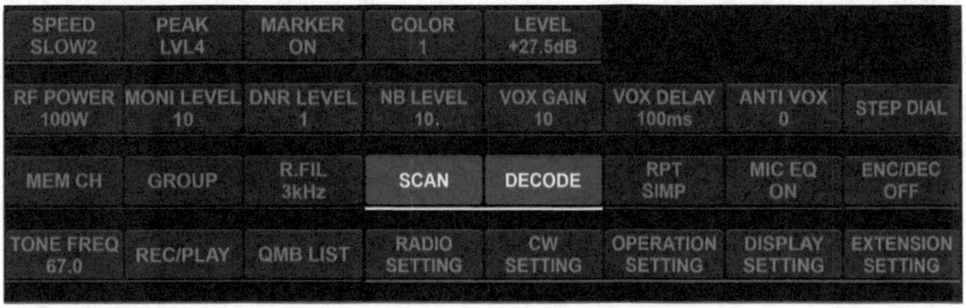

Figure 33: **The FUNC menu – green group**

Item	Function	Behavior
SCAN	Starts scanning. See the section on scan operation on page 95.	Screen disappears. The scan starts immediately. If the radio is in memory mode it scans the memory channels or the selected group. Unless the radio is in MT (memory tune) mode. In which case it acts like a VFO scan. If the radio is in PMS mode, SCAN will activate a scan between two pre-set frequencies.
DECODE	Opens the decode window provided the mode is CW, PSK, or RTTY.	The screen disappears but leaves MONI LEVEL associated with the MULTI control. Always overlays the MAIN spectrum and waterfall display. Cannot be selected if the SUB receiver is selected. See note 1 below.

Note 1: It is a bit annoying that you cannot see the main receiver spectrum scope and the decode window at the same time.

If you close the decode window so that you can see the band scope you have to go back through the FUNC menu to turn it back on.

TIP: You can stop most scans by touching the touchscreen or pressing FUNC. For the 5 MHz 60m band scan you have to touch the waterfall area below the spectrum display or press FUNC.

TIP: *You can shift the decoder to the sub-receiver using <FUNC> <OPERATION SETTING> <GENERAL> <SUB> (the default is Main).*

GROUP 4: RED: RPT AND MIC EQ

This group contains a Soft Key to start to set the repeater offset mode and one that enables the microphone parametric equalizer.

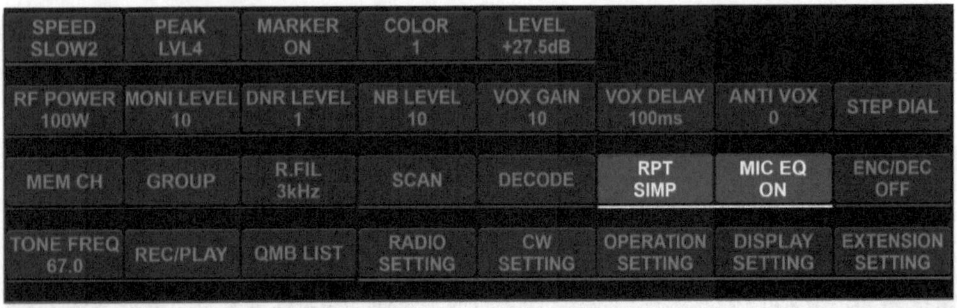

SPEED SLOW2	PEAK LVL4	MARKER ON	COLOR 1	LEVEL +27.5dB				
RF POWER 100W	MONI LEVEL 10	DNR LEVEL 1	NB LEVEL 10	VOX GAIN 10	VOX DELAY 100ms	ANTI VOX 0	STEP DIAL	
MEM CH	GROUP	R.FIL 3kHz	SCAN	DECODE	RPT SIMP	MIC EQ ON	ENC/DEC OFF	
TONE FREQ 67.0	REC/PLAY	QMB LIST	RADIO SETTING	CW SETTING	OPERATION SETTING	DISPLAY SETTING	EXTENSION SETTING	

Figure 34: **The FUNC menu – red group**

Item	Function	Behavior
RPT	Sets the repeater offset for FM operation. You can select SIMP (simplex), + (positive offset), or - (negative offset).	The screen disappears but leaves RPT associated with the MULTI control for about 5 seconds. You have to be very quick to use the MULTI knob to select the offset that you require. The amount of offset for the 10m and 6m bands can be set by menu command, see note 1. The control is only active on FM mode.
MIC EQ	Toggles the parametric microphone equalizer on or off. See note 2 below.	The screen disappears immediately but toggles the parametric microphone equalizer on or off. The control is only active on voice modes.

TIP: *RPT offset.* **Warning!** *Once an offset is applied it stays set for the whole of the band until you change back to simplex. This is pretty horrible, so I suggest you save the repeater channel in a memory slot and then change the offset back to simplex.*

Note 1: The repeater shift for the 28 MHz or 50 MHz band is under,

<FUNC> <RADIO SETTING> <RPT SHIFT (28 MHz)> the default is 100 kHz

<FUNC> <RADIO SETTING> <RPT SHIFT (50 MHz)> the default is 1 MHz

Note 2: The parametric microphone equalizer (EQ) is a three-stage equalizer that can boost audio frequencies by up to 10 dB or cut audio frequencies by up to 20 dB. It is highly configurable. You can adjust the center frequency of each EQ stage, the boost or cut in dB, and the Q or bandwidth of the filter at each stage.

There are two completely different equalizers. One for when the speech processor is on and one for when the speech processor is off. See setting up instructions for the parametric equalizer on page 18.

GROUP 5: BLUE: OTHER SETTINGS

This group seems to contain Soft Keys for items that didn't fit anywhere else.

SPEED SLOW2	PEAK LVL4	MARKER ON	COLOR 1	LEVEL +27.5dB				
RF POWER 100W	MONI LEVEL 10	DNR LEVEL 1	NB LEVEL 10	VOX GAIN 10	VOX DELAY 100ms	ANTI VOX 0	STEP DIAL	
MEM CH	GROUP	R.FIL 3kHz	SCAN	DECODE	RPT SIMP	MIC EQ ON	ENC/DEC OFF	
TONE FREQ 67.0	REC/PLAY	QMB LIST	RADIO SETTING	CW SETTING	OPERATION SETTING	DISPLAY SETTING	EXTENSION SETTING	

Figure 35: **The FUNC menu – blue group**

Item	Function	Behavior
ENC/DEC	Sets tone squelch on FM modes. There are three choices. OFF (default). ENC (encode) applies the CTCSS tone on your transmitted signal to open the repeater squelch. TSQ (tone squelch) sends the tone and also mutes the receiver unless the same tone is received from the repeater.	The screen disappears but leaves ENC/DEC associated with the MULTI control for about 5 seconds. You have to be very quick to use the MULTI knob to select the tone setting that you require. The control is only active on FM and FM-DATA modes. Use TONE FREQ to choose a CTCSS tone frequency.
TONE FREQ	Sets the CTCSS tone frequency for the ENC/DEC function. (67.0 default)	The screen disappears but leaves TONE FREQ associated with the MULTI control for about 5 seconds. You have to be very quick to use the MULTI knob to select the CTCSS tone that you require. The control is only active on FM and FM-DATA modes.

REC/PLAY	Used for accessing the voice messages or the CW, RTTY, or PSK message keyers, depending on the current operating mode.	The FUNC screen disappears and leaves the message memory popup on the screen. You can record or play back the voice or text messages. CW messages can be entered as text or from your CW paddle. See setting up the voice memory keyer (page 20), setting up the CW (page 21), or RTTY and PSK message keyers (page 26).
QMB LIST	Opens a dialogue box for viewing or deleting frequencies stored by the QMB button. See note 1 below.	FUNC screen disappears and leaves the QMB (quick memory button) list active. Press BACK or FUNC to exit.

TIP: ENC/DEC. **Warning!** *Once an encoded tone or tone squelch is enabled it stays set for the whole of that band until you change back to OFF. I suggest you save the repeater channel in a memory slot and then change the ENC/DEC setting back to OFF.*

Note 1: QMB. You can delete a single entry or multiple entries. Touch the line to activate a green tick (check mark) against the frequencies you want to delete. Then touch the DELETE Soft Key. Press BACK or FUNC to exit. I don't see any reason why you would want to delete frequencies in the QMB list unless you saved a frequency by accident. The QMB stores each saved frequency at the top of the list and moves the previous items down, so eventually the oldest ones will drop off the bottom.

There is a menu setting for changing the number of QMB slots from five to ten.

<FUNC> <OPERATION SETTING> <GENERAL> <QMB CH> select 5ch or 10ch

GROUP 6: YELLOW: EQUIPMENT SETTING SUB MENUS

This group provides links to deeper menu layers, such as CW settings, audio parameters, communications parameters, reset, SD card, date and time, and firmware updates. This is only a brief overview. Each of these menu choices will be fully described in the next few chapters.

SPEED SLOW2	PEAK LVL4	MARKER ON	COLOR 1	LEVEL +27.5dB				
RF POWER 100W	MONI LEVEL 10	DNR LEVEL 1	NB LEVEL 10	VOX GAIN 10	VOX DELAY 100ms	ANTI VOX 0	STEP DIAL	
MEM CH	GROUP	R.FIL 3kHz	SCAN	DECODE	RPT SIMP	MIC EQ ON	ENC/DEC OFF	
TONE FREQ 67.0	REC/PLAY	QMB LIST	RADIO SETTING	CW SETTING	OPERATION SETTING	DISPLAY SETTING	EXTENSION SETTING	

Figure 36: **The FUNC menu – yellow group**

Item	Function	Behavior
RADIO SETTING	Settings for the various modes, SSB, AM, FM, PSK, RTTY, and DATA.	Opens a seven-tab setup screen containing mode setup information. See the chapter on Radio Settings on page 125.
CW SETTING	Settings relating to CW operation. Break-in, QSK, audio frequency response, keyer settings, Morse key type, and the CW decoder bandwidth.	Opens the three-tab setup screen containing CW mode setup information. See the chapter on CW Settings on page 137.
OPERATION SETTING	Settings relating to QMB, memory groups, split, language, DSP, Transmitter power, microphone equalizer, VOX, tuning rate and steps, and audio setup information.	Opens a five-tab setup screen containing operational settings, general, RX DSP, transmitter audio, transmitter general, and tuning. See the chapter on Operation Settings on page 142
DISPLAY SETTING	Settings relating to the display, screen saver, mouse speed, spectrum display RBW, bandwidth and sensitivity, and external monitor resolution.	Opens a three-tab setup screen containing the display, scope, and external monitor settings. See the chapter on Display Settings page 151

EXTENSION SETTING	Date and time setup, Save and recall from SD card, current firmware, calibration, and three levels of reset. See note 1.	Opens a five-tab setup screen containing date and time setup, SD card data, current firmware, calibration, and three levels of reset. See the chapter on Extension Settings page 153.

Note 1: Extension Setting Menu. The SOFT VERSION screen is for information only. Firmware updates are performed through the SD Card Firmware Update item.

TIP: Be very careful not to mix up the load and the save buttons in the SD Card sub-menu. Do not format the SD card unless it is a brand-new empty card.

TIP: If you insert the SD card with the transceiver turned on you will be asked if you want to 'SETUP?' If you touch the YES option the SD Card setup menu will open so that you can format a new SD card, save or load the memory channel list, save or load a menu backup, look at the free space and capacity of the SD card, or carry out a firmware update.

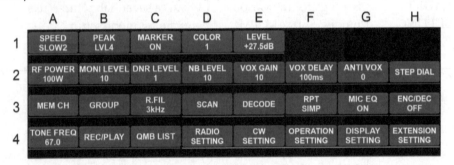

Function	Type	Position
Anti VOX	VOX	G2
Color (waterfall)	Display	D1
CW Setting	Menu	E4
Decoders	Receive	E3
Display Setting	Menu	G4
DNR level	Receive	C2
ENC/DEC (tone)	Repeater	H3
Extension Setting	Menu	H4
Groups	Memory	B3
Level (spectrum)	Display	E1
Markers	Display	C1
MEM CH	Memory	A3
Mic Equalizer	Transmit	G3
Monitor level	Transmit	B2
NB level	Receive	D2

Function	Type	Position
Operation Setting	Menu	F4
Peak (spectrum)	Display	B1
QMB List	Memory	C4
R.FIL roofing filter	Receive	C3
Radio Setting	Menu	D4
REC/PLAY	Transmit	B4
RF Power	Transmit	A2
RPT offset	Repeater	F3
SCAN	Scanning	D3
Speed	Display	A1
Step dial	Tuning	H2
Tone Freq	Repeater	A4
VOX gain	VOX	E2
VOX delay	VOX	F2

Radio settings

The Radio Setting menu, <FUNC> <RADIO SETTING>, contains configuration settings for the SSB, AM, FM, PSK, RTTY, and DATA modes. After making a selection, use the MULTI knob to change the settings. If you make changes to any of the menu settings, it is a good idea to head to <FUNC> <EXTENSION SETTING> <MENU SAVE> <DONE> <NEW> <ENT> to save your changes to the SD card.

SSB MODE

AGC delay: The first three settings control how fast the AGC action will decay after the received signal falls below the AGC threshold if manual AGC has been selected.

TIP: FAST, MID, SLOW, or AUTO AGC can be selected for each band and for each receiver. Touch the AGC Soft Key on the bar that separates the VFO numbers from the spectrum scope or 3DSS display. Yaesu recommends that you leave it set to AUTO which selects the optimum AGC setting for the selected mode.

Receiver audio filter: The next four settings control the shape of the receiver audio passband. I left all these settings at the default settings. LCUT FREQ sets the low-frequency end of the passband and LCUT SLOPE adjusts how sharp the filter is. 6 dB per octave is a soft filter and 18 dB per octave is a sharp filter. You would normally make both ends sharp or soft, not mix them. Naturally, HCUT FREQ sets the high-frequency end of the passband and HCUT SLOPE adjusts how sharp the filter roll-off is at the high end.

SSB receiver output: SSB OUT SELECT determines which receiver will have its audio sent to the RTTY/Data jack on the rear panel of the radio. SSB OUT LEVEL sets the level of the audio signal. These settings are only useful if you are connecting an external device "the old-fashioned way." My connections to the PC are over the single USB cable, so these settings are unimportant.

Transmitter bandwidth: TX BPF SEL selects the bandwidth of the transmitted signal. With the latest firmware, the default for SSB has changed to 100-2900 Hz (2.8 kHz BW). If you are contesting you may prefer a narrow option such as 200-2800 Hz (2.6 kHz), or 300 to 2700 Hz (2.4 kHz).

Audio inputs: SSB MOD SOURCE is a bit "oddball." It can be set to MIC in which case the SSB modulation comes from the microphone and not the RTTY/Data jack on the rear panel or the USB cable. "So far so good." If you set it to REAR, the modulating audio comes from the rear panel if you key the transmitter with the MOX switch. However, the microphone behaves normally if you use the PTT switch on the hand microphone, or key the radio using the PTT jack on the rear panel. I guess that makes sense if you are using the rear panel PTT for a foot switch.

If you did select REAR as the SSB MOD SOURCE, then REAR SELECT sets whether the SSB modulation audio will come from the RTTY/Data jack or the USB cable.

TIP: You would only select REAR if you were planning on sending speech audio to the transceiver from another device. So that the equalizer and compressor settings would be active. You should use the DATA-U or DATA-L modes for connecting external digital mode software. 'That's what they are for!' My advice is to set SSB MOD SOURCE to MIC and REAR SELECT to USB (although that one does not matter).

RPORT GAIN: sets the audio level into the transmitter on SSB mode. Assuming you took my advice and set SSB MOD SOURCE to MIC, then this control is irrelevant. In any case, I would leave it at 50 and adjust the modulation level using the PC (Windows) sound card levels and/or the level control on the external software or connected device.

RPTT SELECT: sets the control line for activating 'keying' the transmitter from the USB cable. Set it to RTS (ready to send), or DAKY if you are using pin 3 of the RTTY/Data jack for PTT.

TIP: The RTS and DTR labels don't matter. You can use either line for the transmit PTT as long as you use the other line for CW or RTTY. The names relate to old-fashioned RS-232 communications between 'old school' computers. RTS stands for 'ready to send' and DTR stands for 'data terminal ready.' But the lines have not been used for that sort of signaling since the 1970s.

RADIO SETTINGS - SSB			
Function	Range	ZL3DW Setting	My Setting
AGC FAST DELAY	20-4000 ms	300 ms (default)	
AGC MID DELAY	20-4000 ms	1000 ms (default)	
AGC SLOW DELAY	20-4000 ms	3000 ms (default)	
LCUT FREQ	Off/100 – 1000 Hz	100 Hz (default)	
LCUT SLOPE	6 or 18 dB/octave	6 dB/oct (default)	
HCUT FREQ	700-4000 Hz	3000 Hz (default)	
HCUT SLOPE	6 or 18 dB/octave	6 dB/oct (default)	
SSB OUT SELECT	MAIN/SUB	MAIN (default)	
SSB OUT LEVEL	0-100	50 (default)	
TX BPF SEL	50-3050, 100-2900, 200-2800, 300-2700, 400-2600 Hz	200-2800 Hz	
SSB MOD SOURCE	MIC/REAR	MIC default)	
REAR SELECT	DATA/USB	USB	
RPORT GAIN	0-100	50 (default)	
RPTT SELECT	DAKY/RTS/DTR	RTS	

AM MODE

AGC delay: The first three settings control how fast the AGC action will decay after the received signal falls below the AGC threshold. FAST, MID, and SLOW refer to the AGC setting that you have selected on the main display.

TIP: The AGC can be selected for each band and for each receiver. It is a Soft Key on the bar that separates the VFO from the spectrum scope or 3DSS display. Yaesu recommends that you leave it set to Auto which makes the AGC decay settings irrelevant.

Receiver audio filter: The next four settings control the shape of the receiver audio passband. I left all these settings at the default settings (OFF). LCUT FREQ sets the low-frequency end of the passband and LCUT SLOPE adjusts how sharp the filter is. 6 dB per octave is a soft filter and 18 dB per octave is a sharp filter. You would normally make both ends sharp or soft, not mix them. Naturally, HCUT FREQ sets the high-frequency end of the passband and HCUT SLOPE adjusts how sharp the filter roll-off is at the high end.

AM receiver output: AM OUT SELECT determines which receiver will have its audio sent to the RTTY/Data jack on the rear panel of the radio. AM OUT LEVEL sets the level of the audio signal. These settings are only useful if you are connecting up an external device "the old-fashioned way." All my connections to the PC are over the single USB cable, so these settings are unimportant.

Transmitter bandwidth: TX BPF SEL selects the bandwidth of the transmitted signal. The default for AM, 50-3050 Hz (3 kHz BW).

Audio inputs: AM MOD SOURCE. This one is a bit "oddball." It can be set to MIC in which case the AM modulation comes from the microphone and not the RTTY/Data jack on the rear panel or the USB cable. "So far so good." If you set it to REAR, the modulating audio comes from the rear panel if you key the transmitter with the MOX switch. However, the microphone behaves normally if you use the PTT switch on the hand microphone, or key the radio using the PTT jack on the rear panel. If you did select REAR as the AM MOD SOURCE, then REAR SELECT sets whether the AM modulation audio will come from the RTTY/Data jack or the USB cable.

TIP: You would only select REAR if you were planning on sending speech audio to the transceiver from another device. So that the equalizer and compressor settings would be active. My advice is to set AM MOD SOURCE to MIC and REAR SELECT to USB.

MIC GAIN: sets the microphone gain for AM. You can set it anywhere from 1 to 100 using the MULTI knob, in which case you should not have to adjust it again unless you change microphones. Or you can leave it on the default MCVR setting. When it is set to MCVR the AM microphone gain is adjustable by rotating the MIC/SPEED control on the front panel.

Changing the MIC GAIN affects the microphone level for SSB, AM, and FM. I recommend setting it for SSB and then leaving it alone. If you want a different microphone sensitivity for AM, use one of the 100 pre-set levels.

RPORT GAIN: sets the audio level into the transmitter from the USB port or RTTY/Data jack on AM mode. I would leave it at 50 and adjust the modulation level using the PC (Windows) sound card levels and/or the level control on the external software or connected device.

RPTT SELECT: sets the control line for activating 'keying' the transmitter from the USB cable. Set it to RTS (ready to send), or DAKY if you are using pin 3 of the RTTY/Data jack for PTT.

TIP: The RTS and DTR labels don't matter. You can use either line for the transmit PTT as long as you use the other line for CW or RTTY. The names relate to old-fashioned RS-232 communications between 'old school' computers. RTS stands for 'ready to send' and DTR stands for 'data terminal ready.' But the lines have not been used for that sort of signaling since the 1970s. Anyway, I always use 'ready to send' for the 'Send' or 'PTT' command and 'data terminal ready' for the CW or RTTY data signal.

RADIO SETTINGS - AM			
Function	Range	ZL3DW Setting	My Setting
AGC FAST DELAY	20-4000 ms	1000 ms (default)	
AGC MID DELAY	20-4000 ms	2000 ms (default)	
AGC SLOW DELAY	20-4000 ms	4000 ms (default)	
LCUT FREQ	Off/100 – 1000 Hz	OFF (default)	
LCUT SLOPE	6 or 18 dB/octave	6 dB/oct (default)	
HCUT FREQ	700-4000 Hz	OFF (default)	
HCUT SLOPE	6 or 18 dB/octave	6 dB/oct (default)	
AM OUT SEL	MAIN/SUB	MAIN (default)	
AM OUT LVL	0-100	50 (default)	
TX BPF SEL	50-3050, 100-2900, 200-2800, 300-2700, 400-2600 Hz	50-3050 Hz (default)	
AM MOD SOURCE	MIC/REAR	MIC (default)	
MIC GAIN	MCVR/0-100	MCVR (default)	
REAR SELECT	DATA/USB	USB	
RPORT GAIN	0-100	50 (default)	
RPTT SEL	DAKY/RTS/DTR	DAKY (default)	

FM MODE

AGC delay: The first three settings control how fast the AGC action will decay after the received signal falls below the AGC threshold. FAST, MID, and SLOW refer to the manual AGC setting that you have selected on the main display.

TIP: The AGC can be selected for each band and for each receiver. It is a Soft Key on the bar that separates the VFO from the spectrum scope or 3DSS display. Yaesu recommends that you leave it set to Auto which makes the AGC decay settings irrelevant.

Receiver audio filter: The next four settings control the shape of the receiver audio passband. I left all these settings at the default settings. LCUT FREQ sets the low-frequency end of the passband and LCUT SLOPE adjusts how sharp the filter is. 6 dB per octave is a soft filter and 18 dB per octave is a sharp filter. The default for FM is to use the sharp filter. You would normally make both ends sharp or soft, not mix them. HCUT FREQ sets the high-frequency end of the passband and HCUT SLOPE adjusts how sharp the filter roll-off is at the high end.

FM receiver output: FM OUT SELECT determines which receiver will have its audio sent to the RTTY/Data jack on the rear panel of the radio. FM OUT LEVEL sets the level of the audio signal. These settings are only useful if you are connecting up an external device "the old-fashioned way." All my connections to the PC are over the single USB cable, so these settings are unimportant.

Audio inputs: FM MOD SOURCE. This one is a bit "oddball." It can be set to MIC in which case the FM modulation comes from the microphone and not the RTTY/Data jack on the rear panel or the USB cable. "So far so good." If you set it to REAR, the modulating audio comes from the rear panel if you key the transmitter with the MOX switch. However, the microphone behaves normally if you use the PTT switch on the hand microphone, or key the radio using the PTT jack on the rear panel. If you did select REAR as the FM MOD SOURCE, then REAR SELECT sets whether the AM modulation audio will come from the RTTY/Data jack or the USB cable.

TIP: You would only select REAR if you were planning on sending speech audio to the transceiver from another device. So that the equalizer and compressor settings would be active. My advice is to set FM MOD SOURCE to MIC and REAR SELECT to USB.

MIC GAIN: sets the microphone gain for FM. You can set it anywhere from 1 to 100 using the MULTI knob, in which case you should not have to adjust it again unless you change microphones. Or you can leave it on the default MCVR setting. When it is set to MCVR the FM microphone gain is adjustable by rotating the MIC/SPEED control on the front panel. Changing the MIC GAIN affects the microphone level for SSB, AM, and FM. I recommend setting it for SSB and then leaving it alone. If you want a different microphone sensitivity for FM, use one of the 100 pre-set levels.

RPORT GAIN: sets the audio level into the transmitter from the USB port or RTTY/Data jack on AM mode. I would leave it at 50 and adjust the modulation level using the PC (Windows) sound card levels and/or the level control on the external software or connected device.

RPTT SELECT: sets the control line for activating 'keying' the transmitter from the USB cable. Set it to RTS (ready to send), or DAKY if you are using pin 3 of the RTTY/Data jack for PTT.

Repeater offsets: You can use RPT SHIFT (28 MHz) to change the offset between your receiver and transmitter frequencies for operating through an FM repeater on the 10m band. The default 100 kHz offset suits most repeaters in most countries.

You can use RPT SHIFT (50 MHz) to change the offset between your receiver and transmitter frequencies for operating through an FM repeater on the 6m band. The default 1 MHz offset suits the repeaters in most countries. But some 6m repeaters use a 500 kHz offset.

Note that this setting does not allow you to set whether the repeater is using a 'plus' or a 'minus' offset. To do that, you have to use the RPT Soft Key described in 'Group 4' in the previous chapter.

RADIO SETTINGS - FM			
Function	Range	ZL3DW Setting	My Setting
AGC FAST DELAY	20-4000 ms	160 ms (default)	
AGC MID DELAY	20-4000 ms	500 ms (default)	
AGC SLOW DELAY	20-4000 ms	1500 ms (default)	
LCUT FREQ	Off/100 – 1000 Hz	300 Hz (default)	
LCUT SLOPE	6 or 18 dB/octave	18 dB/oct (default)	
HCUT FREQ	700-4000 Hz	3000 Hz (default)	
HCUT SLOPE	6 or 18 dB/octave	18 dB/oct (default)	
FM OUT SEL	MAIN/SUB	MAIN (default)	
FM OUT LVL	0-100	50 (default)	
FM MOD SOURCE	MIC/REAR	MIC (default)	
MIC GAIN	MCVR/0-100	MCVR (default)	
REAR SEL	DATA/USB	USB	
RPORT GAIN	0-100	50 (default)	
RPTT SEL	DAKY/RTS/DTR	RTS	
RPT SHIFT (28 MHz)	0-1000 kHz	100 kHz (default)	
RPT SHIFT (50 MHz)	0-4000 kHz	1000 kHz (default)	

PSK AND DATA MODES

These settings are for the PSK mode and the DATA modes. Most digital mode software expects that you will use the upper sideband DATA-U mode on all bands.

AGC delay: The first three settings control how fast the AGC action will decay after the received signal falls below the AGC threshold. FAST, MID, and SLOW refer to the AGC setting that you have selected on the main display. The default settings are the same as for FM.

TIP: The AGC can be selected for each band and for each receiver. It is a Soft Key on the bar that separates the VFO from the spectrum scope or 3DSS display. Yaesu recommends that you leave it set to Auto which makes the AGC decay settings irrelevant.

PSK setting: The manual does not really say what PSK TONE does. It sets the carrier offset for a PSK signal when the radio is in the PSK mode. It is rather like the CW pitch setting. At the displayed VFO frequency a received PSK signal will sound like a 1 kHz (default), 1.5 kHz, or 2 kHz signal.

The **DATA SHIFT (SSB)** setting does the same thing for the DATA_U or DATA-L modes. It sets the frequency offset above (or below) the carrier frequency that a data signal will be placed. It is the offset that will apply for FSK RTTY or 300 baud HF packet radio. You can adjust the offset from 0 to 3 kHz in 10 Hz steps. The default is 1500 Hz. I can't see any reason to change it.

Receiver audio filter: The next four settings control the shape of the receiver audio passband. I left all these settings at the default settings. LCUT FREQ sets the low-frequency end of the passband and LCUT SLOPE adjusts how sharp the filter is. 6 dB per octave is a soft filter and 18 dB per octave is a sharp filter. The default for PSK/DATA is to use the sharp filter. You would normally make both ends sharp or soft, not mix them. Naturally, HCUT FREQ sets the high-frequency end of the passband and HCUT SLOPE adjusts how sharp the filter roll-off is at the high end.

PSK or DATA receiver output: DATA OUT SELECT determines which receiver will have its audio sent to the RTTY/Data jack on the rear panel of the radio. DATA OUT LEVEL sets the level of the audio signal. These settings are only useful if you are connecting an external device "the old-fashioned way." My connections to the PC are over the single USB cable, so these settings are unimportant.

Transmitter bandwidth: TX BPF SEL selects the bandwidth of the transmitted signal. I changed the DATA mode bandwidth to 100-2900 Hz (2.8 kHz BW) because it almost matches the bandwidth of the 200-3000 Hz receiver passband displayed on the digital mode software. The 50-3050 Hz option may fit better, but I would not be comfortable transmitting a digital mode signal below 100 Hz.

Audio inputs: DATA MOD SOURCE. Needs to be set to REAR so that you can send data from an external digital modes program over the USB cable. REAR

SELECT should be set to USB unless you are using a Rig Runner or old-fashioned audio cables.

RPORT GAIN: sets the audio level into the transmitter from the USB port or RTTY/Data jack on DATA mode. I would leave it at 50 and adjust the modulation level using the PC (Windows) sound card levels and/or the level control on the external software or connected device.

RPTT SELECT: sets the control line for activating 'keying' the transmitter from the USB cable. Set it to RTS (ready to send), or DAKY if you are using pin 3 of the RTTY/Data jack for PTT.

	RADIO SETTINGS – PSK and DATA		
Function	Range	ZL3DW Setting	My Setting
AGC FAST DELAY	20-4000 ms	160 ms (default)	
AGC MID DELAY	20-4000 ms	500 ms (default)	
AGC SLOW DELAY	20-4000 ms	1500 ms (default)	
PSK TONE	1000/1500/2000	1000 Hz (default)	
DATA SHIFT (SSB)	0-3000 Hz	1500 Hz (default)	
LCUT FREQ	Off/100 – 1000 Hz	300 Hz (default)	
LCUT SLOPE	6 or 18 dB/octave	18 dB/oct (default)	
HCUT FREQ	700-4000 Hz	3000 Hz (default)	
HCUT SLOPE	6 or 18 dB/octave	18 dB/oct (default)	
DATA OUT SELECT	MAIN/SUB	MAIN (default)	
DATA OUT LEVEL	0-100	50 (default)	
TX BPF SEL	50-3050, 100-2900, 200-2800, 300-2700, 400-2600 Hz	100-2900 Hz	
DATA MOD SOURCE	MIC/REAR	REAR	
REAR SEL	DATA/USB	USB	
RPORT GAIN	0-100	50 (default)	
RPTT SEL	DAKY/RTS/DTR	RTS	

RTTY MODE

These settings are for the internal RTTY mode. The normal mode for amateur radio RTTY is lower sideband RTTY-L for all bands.

RTTY is a frequency shift keying (FSK) mode. It works by sending a tone on a 'mark' frequency and then applying frequency shifts to a 'space' frequency to send the Baudot code. For amateur radio RTTY, the 'space' frequency is usually 170 Hz lower than the 'mark' frequency. The mark frequency is the frequency indicated by the VFO. It has been offset from the carrier point frequency by the **Mark Frequency** usually 2125 Hz.

AGC delay: The first three settings control how fast the AGC action will decay after the received signal falls below the AGC threshold. FAST, MID, and SLOW refer to the AGC setting that you have selected on the main display. The default settings are the same as the ones chosen for FM, PSK, and DATA.

RTTY polarity: POLARITY RX and POLARITY TX invert the tone frequencies of the RTTY signal. Amateur radio RTTY uses the normal (NOR) mode where the 'space' frequency is lower than the 'mark' frequency. Commercial operators on the shortwave bands often use reverse (REV) mode where the 'space' frequency is higher than the 'mark' frequency.

Receiver audio filter: The next four settings control the shape of the receiver audio passband. I left all these settings at the default settings. LCUT FREQ sets the low-frequency end of the passband and LCUT SLOPE adjusts how sharp the filter is. 6 dB per octave is a soft filter and 18 dB per octave is a sharp filter. The default for RTTY is to use the sharp filter. You would normally make both ends sharp or soft, not mix them. Naturally, HCUT FREQ sets the high-frequency end of the passband and HCUT SLOPE adjusts how sharp the filter roll-off is at the high end.

RTTY receiver output: RTTY OUT SELECT determines which receiver will have its audio sent to the RTTY/Data jack on the rear panel of the radio. RTTY OUT LEVEL sets the level of the audio signal. These settings are only useful if you are connecting an external device "the old-fashioned way." My connections to the PC are over the single USB cable, so these settings are unimportant.

RPTT SELECT: sets the control line for activating 'keying' the transmitter from the USB cable. Set it to RTS (ready to send), or DAKY if you are using pin 3 of the RTTY/Data jack for PTT.

RTTY mark and shift: MARK FREQUENCY sets the audio tone for the mark frequency. The choices are the two internationally recognized 'low tones.' The default 2125 Hz setting is the most used offset, but the alternative setting of 1275 Hz is perfectly acceptable as well. The SHIFT frequency for amateur radio is 170 Hz.

The other shift choices are for receiving commercial RTTY transmissions which may use other shift frequencies and baud rates.

RADIO SETTINGS - RTTY			
Function	Range	ZL3DW Setting	My Setting
AGC FAST DELAY	20-4000 ms	160 ms (default)	
AGC MID DELAY	20-4000 ms	500 ms (default)	
AGC SLOW DELAY	20-4000 ms	1500 ms (default)	
POLARITY RX	NOR/REV	NOR (default)	
POLARITY TX	NOR/REV	NOR (default)	
LCUT FREQ	1000/1500/2000	300 Hz (default)	
LCUT SLOPE	6 or 18 dB/octave	18 dB/oct (default)	
HCUT FREQ	700-4000 Hz	3000 Hz (default)	
HCUT SLOPE	6 or 18 dB/octave	18 dB/oct (default)	
RTTY OUT SEL	MAIN/SUB	MAIN (default)	
RTTY OUT LVL	0-100	50 (default)	
RPTT SEL	DAKY/RTS/DTR	RTS	
MARK FREQUENCY	1275 or 2125	2125 Hz (default)	
SHIFT FREQUENCY	170, 200, 425 or 850 Hz	170 Hz (default)	

ENCODING AND DECODING PSK

You can choose from BPSK which is standard PSK-31, or QPSK. I assume that it is QPSK-31 since there is no baud rate control. QPSK has the advantage of having built-in error correction but it is considerably less sensitive than BPSK and much less popular on the bands. The baud rate of 31 symbols per second of PSK-31 is chosen to provide a narrow bandwidth transmission at a typical typing speed. These days there is little PSK activity since most of the world has moved to FT8.

The internal PSK decoder has AFC (automatic frequency control) to help you tune the PSK signal accurately. It will pull the receiver slightly to get the signal tuned in. The DECODE AFC RANGE sets the bandwidth for the AFC action. A BPSK-31 signal has a bandwidth of about 31 Hz. So, I would experiment with using the 30 Hz or 15 Hz settings

QPSK phase shift: You can change the direction of the QPSK phase shift using the QPSK POLARITY settings. If you can't decode a QPSK signal, it is possible that the other station is using the other mode.

The **PSK TX LEVEL** can be set from 0 to 100. The default setting is 70. I expect that this limits the transmitted power to 70% of maximum power.

Which is a good thing for modes that continuously run at full power like PSK. I would leave it at 70 unless you have a good reason to change it. Running PSK at 100% power could overheat your transmitter.

RADIO SETTINGS – ENCODING & DECODING PSK			
Function	Range	ZL3DW Setting	My Setting
PSK MODE	BPSK or QPSK	BPSK (default)	
DECODE AFC RANGE	8, 15, or 30 Hz	15 Hz (default)	
QPSK POLARITY RX	NOR/REV	NOR (default)	
QPSK POLARITY TX	NOR/REV	NOR (default)	
PSK TX LEVEL	0-100	70 (default)	

ENCODING AND DECODING RTTY

These settings relate to FSK RTTY generated by the radio in the RTTY mode. You can also operate RTTY in the DATA-U mode using AFSK RTTY from an external digital modes program.

[FSK = frequency shift keying], [AFSK = audio frequency shift keying].

USOS: stands for 'unshift on space.' The Baudot code used to send RTTY dates back to the old mechanical teletype machine days. It is only a 5-bit code. Five bits only allows for 32 different characters, so there are not enough combinations to encode all the upper-case letters and numerals 0 to 9. Baudot gets around this by sending a special 'FIGS' code before sending numbers and then sending a 'LTRS' code to switch the decoder back to letters when the numbers have been sent. Problems arise if noise on the signal prevents the FIGS or LTRS code from being received properly. If that happens the decoder outputs a string of garbage characters until the next time a FIGS or LTRS code arrives. If you have TX USOS turned on, the encoder will send a LTRS code every time a 'space' character is sent. Of course, this usually happens at the end of every word. These regular reminders that the data is letters, not numbers, greatly improves the data synchronization between the transmitting station and the receiving station, resulting in fewer errors in the text. If you have RX USOS turned on, the decoder in the radio will return to letters mode every time a space character is received regardless of whether LTRS codes are sent from the distant station.

The moral of the story is, leave RX USOS and TX USOS turned on.

RX NEW LINE CODE: The decoded RTTY text is much easier to read if the lines and paragraphs don't run together. This setting starts a new line on the decode screen each time a CR (carriage return), LF (line feed), or combined CR+LF character is received. You can set it to only inserting a new line when the combined CR+LF character is received. But I recommend leaving it at the default setting.

TX DIDDLE: When switched to transmit mode, mechanical RTTY machines would send a blank signal at any time that a character was not being sent. It is called a 'diddle' because it sounds like a warble, mostly mark tone with regular transitions to the space frequency. It allowed the receiver at the distant end to stay locked.

If you are a slow typist or you pause between words, the RTTY encoder software will send the diddle signal between words or characters as well. Yaesu has given you the option of sending the BLANK signal, or a string of LTRS codes to ensure that the decoder at the distant station is set for receiving letters, or you can turn the diddle off. In which case, nothing is sent except the text message. If signals are reasonable, that should be fine. The only reason to send the diddle is to keep the distant station synchronized and tuned to your signal. The default is BLANK.

The final setting is the BAUDOT CODE. This one is a mystery. I have never seen this choice offered before. I believe that the US setting refers to the ITA2 (international telegraph alphabet no 2) alphabet and the CCITT version refers to the 'Murray' code, otherwise known as CCITT2. Apparently, there are very minor differences in the way some of the letters and characters look when printed.

RADIO SETTINGS – ENCODING & DECODING RTTY			
Function	Range	ZL3DW Setting	My Setting
RX USOS	OFF/ON	ON (default)	
TX USOS	OFF/ON	ON (default)	
RX NEW LINE CODE	CR, LF, CR+LF or CR+LF	CR, LF, CR+LF (default)	
TX DIDDLE	OFF, BLANK, or LTRS	BLANK (default)	
BAUDOT CODE	CCITT or US	US (default)	

CW Settings

The CW Setting menu, <FUNC> <CW SETTING>, includes settings relating to CW operation. Break-in, QSK, audio frequency response, keyer settings, Morse key type, and the CW decoder bandwidth.

TIP: If you make changes to any of the menu settings, it is a good idea to head to <FUNC> <EXTENSION SETTING> <MENU SAVE> <DONE> <NEW> <ENT> and save your changes to the SD card.

CW MODE

AGC delay: The first three settings control how fast the AGC action will decay after the received signal falls below the AGC threshold. FAST, MID, and SLOW refer to the AGC setting that you have selected on the main display.

TIP: The AGC can be selected for each band and for each receiver. It is a Soft Key on the bar that separates the VFO from the spectrum scope or 3DSS display. Yaesu recommends that you leave it set to Auto which makes the AGC decay settings irrelevant.

Receiver audio filter: The next four settings control the shape of the receiver audio passband. I left all these settings at the default settings. LCUT FREQ sets the low-frequency end of the passband and LCUT SLOPE adjusts how sharp the filter is. 6 dB per octave is a soft filter and 18 dB per octave is a sharp filter. The default for CW is to use the sharp filter. You would normally make both ends sharp or soft, not mix them. HCUT FREQ sets the high-frequency end of the passband and HCUT SLOPE adjusts how sharp the filter roll-off is at the high end. The HCUT frequency is lower than on the other modes because you usually use the narrow 600 Hz roofing filter for CW. Or possibly the 300 Hz roofing filter on the FTDX101MP.

CW receiver output: CW OUT SELECT determines which receiver will have its audio sent to the RTTY/Data jack on the rear panel of the radio. CW OUT LEVEL sets the level of the audio signal. These settings are only useful if you are connecting up an external device "the old-fashioned way." My connections to the PC are over the single USB cable, so these settings are unimportant.

CW AUTO MODE: This lets you enable or disable the Morse key or paddle when you are on SSB. Some people like to use a CW sign-off when operating SSB and there is certainly no prohibition against using CW on the SSB band segment. When set to ON the key is 'live' but CW will only be transmitted if the BK-IN switch is enabled (LED lit). When set to OFF, the key is disabled on SSB. If you select 50M the key is active on the 6m band but not on the HF bands. Unless you really want the ability to send CW while in SSB mode, I suggest leaving this function turned off.

TIP: You might consider leaving the CW AUTO MODE set to ON and controlling whether the CW will be transmitted by activating the BK-IN switch. This is fine. But note that BK-IN *also has to be turned on for the five-message voice keyer to work.*

CW BREAK-IN TYPE: selects semi break-in or full break-in.

CW BREAK-IN DELAY: sets the delay before the transceiver switches back to receive when operating in semi break-in mode. It switches back immediately if you are operating in full break-in mode. The setting will depend on your operating style, how long you pause between characters or words, and the speed that you are sending. You will probably have to experiment with the delay setting to stop the radio switching too early. The break-in delay is adjustable from 30 ms to 3 seconds. The three-second setting would hold the radio on transmit between sentences. I changed this setting to 500 ms so that the radio does not switch to receive between words when sending a text macro at 20 wpm.

QSK DELAY TIME: sets the delay before the transmitter starts to transmit the CW signal. This time is important when you are using a linear amplifier. It must be long enough for the amplifier to switch from receive to transmit before RF power arrives. If it is too short the amplifier will be 'hot switching' which is very bad for the relay contacts. It could also truncate the first dit. If you have an old valve (tube) amplifier it is likely that the default setting of 15 ms will not be enough. You should increase the delay and avoid full break-in operation. If your amplifier is tripping every time you send a CW signal try increasing the QSK DELAY TIME. Note that if the keyer is set at 45 wpm or higher the QSK DELAY TIME will be 15 ms regardless of the setting you have selected.

CW RISE TIME: sets the shape of the CW waveform (keying envelope). The default setting is 4 ms. Selecting a slower rise time will make your signal sound softer. Setting fast rise times will make your signal sound harsh. Fast rise times should only be selected if you are using high-speed CW.

CW VFO frequency display: The CW FREQ DISPLAY setting adjusts the frequency offset that will be applied to the VFO when you tune to a CW signal and then change from SSB to CW or CW to SSB.

The default PITCH OFFSET setting offsets the VFO when you switch from CW to SSB, so you hear the CW signal at the same tone. However, if you select the DIRECT FREQUENCY setting the VFO frequency will not be offset. The CW tone will be at the zero-beat frequency and will not be heard when you change to SSB. I believe that the default PITCH OFFSET setting is preferable. It means that you can tune in a CW signal while on SSB and then switch to CW without losing the signal.

CW signals are received at an offset equal to the 'pitch' setting of the PROC/PITCH control. I set my pitch control to 700 Hz.

PC KEYING: sets the control line for 'CW keying' the transmitter from the USB cable. Assuming that you are using RTS for PTT, set it to DTR (device terminal ready). Set it to DAKY if you are using pin 3 of the RTTY/Data jack for PTT.

TIP: If you are sending CW from an external device the BK-IN button does not need to be turned on for the signal to be transmitted.

TIP: The CW keying signal is sent via the DTR line on the 'Standard' COM port, not the 'Enhanced' COM port you are using for CAT.

CW INDICATOR: This setting turns on, or off, the CW tuning indicator below the DSP filter function display. The tuning guide is handy for tuning a CW signal exactly on frequency. However, you can turn it off if you find it distracting.

CW SETTINGS – CW MODE			
Function	Range	ZL3DW Setting	My Setting
AGC FAST DELAY	20-4000 ms	160 ms (default)	
AGC MID DELAY	20-4000 ms	500 ms (default)	
AGC SLOW DELAY	20-4000 ms	1500 ms (default)	
LCUT FREQ	Off/100 – 1000 Hz	250 Hz (default)	
LCUT SLOPE	6 or 18 dB/octave	18 dB/oct (default)	
HCUT FREQ	700-4000 Hz	1200 Hz (default)	
HCUT SLOPE	6 or 18 dB/octave	18 dB/oct (default)	
CW OUT SEL	MAIN/SUB	MAIN (default)	
CW OUT LVL	0-100	50 (default)	
CW AUTO MODE	OFF/50M/ON	OFF (default)	
CW BK-IN TYPE	SEMI/FULL	SEMI (default)	
CW BREAK-IN DELAY	30-3000 ms	200 ms (default)	
CW WAVE SHAPE	1/2/4/6 ms	6 ms (default)	
CW FREQ DISPLAY	DIRECT/PITCH	PITCH OFFSET	
PC KEYING	OFF/DAKY/RTS/DTR	DTR	
QSK DELAY TIME	15/20/25/30 ms	15 ms (default)	
CW INDICATOR	ON/OFF	ON (default)	

KEYER

The Keyer sub-menu sets the parameters of your Morse key or paddle. You can use a 'straight key, a 'Bug,' or a 'Paddle.' There are the usual Iambic choices and an ACS mode which automatically controls the spacing, improving the quality of your Morse. I expect it takes some getting used to.

The 'F' prefix indicates the front panel jack and 'R' indicates the rear panel jack.

KEYER TYPE: sets the type of Morse key. You can have a different type plugged into the back of the radio.

- OFF is for a straight key. The dots contact (Tip) is used. (Or you can just turn the Keyer off).
- BUG uses the keyer to send the dots but lets you time the dashes.
- ELEKEY-A (Iambic mode A). The keyer finishes sending the last symbol, a dot or dash, and stops. If you are sending a dot and you press both sides of the paddle then let go, the keyer should send an 'A.' If you are sending a dash and press both sides of the paddle then let go, the keyer should send an 'N.'
- ELEKEY-B (Iambic mode B). The keyer finishes sending the last symbol, a dot or dash, and then sends the opposite symbol. If you are sending a dot and you press both sides of the paddle then let go, the keyer should send an 'R.' If you are sending a dash and press both sides of the paddle then let go, the keyer should send a 'K.' Mode B is quite good for sending CQ.
- ELEKEY-Y (Iambic mode Y). Is a variant on mode B. If it is transmitting a dash, the keyer ignores the first dot when you squeeze the paddles.
- ACS makes sure that the character spacing remains correct. Forcing a one dash space between characters.

KEYER DOT/DASH: reverses the dots and dashes on a Bug or Paddle.

- NOR (normal) sets the dots to the left paddle and dashes to the right paddle.
- REV (reverse) sets the dashes to the left paddle and dots to the right paddle.

NOTE: The Yaesu manual has this worded incorrectly.

CW WEIGHT: selects the ratio of the length of a dash compared to a dot. The standard is that a dash is three times as long as a dot, but you can set it anywhere from 2.5 to 4.5 times.

NUMBER STYLE: Is used to format the contest number in CW message macros. If you don't use the macros, or you don't operate in CW contests you can ignore this setting.

- 1290 sends the contest numbers using standard Morse code.
- AUNO sends A instead of one, U instead of two, N for nine, and O for zero.
- AUNT sends A instead of one, U instead of two, N for nine, and T for zero.
- A2NO sends A instead of one, N for nine, and O for zero.
- A2NT sends A instead of one, N for nine, and T for zero.

- 12NO sends N for nine, and O for zero. Other numbers are sent normally.
- 12NT sends N for nine, and T for zero. Other numbers are sent normally.

CONTEST NUMBER: If you are using the CW macros in a contest. Set this to 1 before the contest starts. It will increment as the contest numbers are given out. It is only used to format the contest number in CW message macros. You can decrement the contest number using the DEC key on the FH-2 keypad or the REC/PLAY CW Keyer.

CW MEMORY 1-5: You can load messages into the CW keyer in two ways, but not from this screen. Why? I don't know. It would have made sense to be able to touch and hold a Soft Key and edit the contents. If the memory slot is set to TEXT you can type in a message on the on-display keyboard. If the memory slot is set to MESSAGE, you can use the CW paddle or key to save a message to the CW keyer.

REPEAT INTERVAL: If you touch and hold any of the message memories displayed in the REC/PLAY dialogue box, the message will repeat at intervals set by this menu item, until you stop the message by sending a dot or dash from the paddle or touching the message memory key again.

CW SETTINGS – KEYER			
Function	Range	ZL3DW Setting	My Setting
F KEYER TYPE	See keyer type above	ELEKEY-A	
F KEYER DOT/DASH	NOR/REV	NOR (default)	
R KEYER TYPE	See keyer type above	ELEKEY-A	
R KEYER DOT/DASH	NOR/REV	NOR (default)	
CW WEIGHT	2.5-3.5	3.0 (default)	
NUMBER STYLE	See number style	1290 (default)	
CONTEST NUMBER	1-9999	1 (default)	
CW MEMORY 1-5	See CW memory	TEXT	
REPEAT INTERVAL	1-60 seconds	5 sec (default)	

DECODE CW

The internal CW decoder has AFC (automatic frequency control) to help you tune the CW signal accurately. It will pull the receiver slightly to get the signal tuned in. The DECODE CW adjustment sets the bandwidth for the AFC action. The risk in using the wide 250 Hz setting is that in a busy contest it might lock to the wrong CW signal. I plan to stay with the default 100 Hz setting unless I need to change it for some reason.

Operation settings

The Operation Setting menu, <FUNC> <OPERATION SETTING>, has settings relating to QMB, memory groups, split, language, DSP, transmitter power, microphone equalizer, VOX, tuning rate and steps, and audio setup.

TIP: If you make changes to any of the menu settings, it is a good idea to head to <FUNC> <EXTENSION SETTING> <MENU SAVE> <DONE> <NEW> <ENT> and save your changes to the SD card.

GENERAL SETTINGS

<FUNC> <OPERATION SETTING> <GENERAL>

Decode RX select allows you to choose which receiver the CW, PSK, and RTTY decoder will use. MAIN or SUB.

Headphone Mix is a useful setting. It changes the mix of main and sub-receiver audio signals into the headphones.

> SEPARATE – the main receiver is heard on the left channel and the sub-receiver is heard on the right channel.
>
> COMBINE-1 – this is a great option. It places the main receiver on the left channel but mixes in the sub-receiver at a lower level. On the right channel, you hear the sub-receiver with the main receiver at a lower level.
>
> COMBINE-2 – This is the mono option. The audio from both receivers is heard from both channels.

ANT3 select changes the function of the ANT 3 antenna connector. You can choose TRX, R3-T1, R3-T2, or RX-ANT.

- TRX makes ANT3 the same as the other two antenna ports. The radio will transmit into and receive from the antenna connected to the ANT3 connector. If selected, the ANT Soft key will display 3.
- With the R3-T1 option, the transceiver will transmit into the antenna connected to ANT1 and receive from the antenna connected to ANT3. If selected, the ANT Soft key will display R/T1.
- With the R3-T2 option, the transceiver will transmit into the antenna connected to ANT2 and receive from the antenna connected to ANT3. If selected, the ANT Soft key will display R/T2.
- With the RX-ANT option, transmit operation is disabled and the transceiver will receive from the antenna connected to ANT3. If selected, the ANT Soft key will display RANT.

NB width changes the duration that the receiver audio will remain attenuated after the noise blanker responds to a noise spike. It can be set to 1, 3, or 10 ms. The short settings can be used for sharp pulse noise such as an electric fence or car ignition noise. The longer setting is better suited to longer nose bursts such as lightning.

NB rejection sets the depth of noise blanker attenuation. You can set 10, 30, or 40 dB of attenuation. The 30 dB default setting should be fine most of the time. If the interference is not too loud a 10 dB setting may be adequate. The DSP noise reduction algorithm looks forward in the digital data stream representing the receiver audio. This introduces a very small delay (latency) to the signal. When the noise level triggers a response, it attenuates the audio for a period set by the NB width control.

TIP: The noise blanker (NB) level is the threshold level at which the noise blanker will respond to noise spikes. You can assign it to the MULTI knob for a few seconds by press and holding the NB button. But it is better to press and hold the CS button and assign NB LEVEL to the big VFO tuning ring.

TIP: Most DSP noise blankers (NB) work by eliminating or attenuating noise peaks that are above the average received signal level. They usually have no effect on noise pulses that are below the average speech level. The noise blanker is best used to reduce or eliminate regular pulse-type noise such as car ignition noise. You may need to experiment with the NB rejection level and the NB width controls when tackling a particular noise problem. If the noise blanker has no noticeable effect, try the DNR (digital noise reduction) button.

Beep level sets how loud the beeps are. Set (0-100), the default is 10.

RF/SQL VR sets the function of the RF/SQL control. The outer knob controls either the RF gain of the receiver or the squelch. The default setting for the RF/SQL control is for it to act as an RF Gain control. The RF Gain control provides a manual adjustment of the gain of the RF and I.F. stages. It can be very useful to turn down the RF gain to improve the signal to noise ratio of reasonably strong signals, especially on the noisy lower frequency bands.

But most people leave it turned up to maximum all the time. In fact, the Yaesu manual states that the "RF/SQ knob is normally left in the fully clockwise position." I find it much more useful to change the menu setting so that the control operates the receiver squelch. If you operate mostly on 40m, 80m, and 160m I recommend setting the menu to the RF Gain position. If like me you favor the higher bands, set the menu to SQL. This menu setting affects both receivers.

TIP: I reversed the RF/SQL knob on both receivers. Gently and slowly pull the volume and squelch knobs off and reinstall the RF/SQL knob 180 degrees offset. It makes the squelch work backwards, but it stops the lever on the main receiver knob from getting in the way of the all-important MULTI knob.

Tuner select is used to tell the radio if an external Yaesu compatible antenna tuner is connected to one of the antenna connectors. INT means the internal antenna is in use. EXT1, EXT2, and EXT3 are used to indicate that an external antenna tuner is connected to ANT1, ANT2, or ANT3 respectively. This setting is only for Yaesu compatible antenna tuners.

232C RATE and 232C TIME OUT TIMER: These two settings relate to the serial port settings for the 9 pin DB-9 serial port on the back of the radio. They do not affect the USB connection.

CAT settings: CAT RATE is used to set the interface speed of the USB connection between the radio and the PC software. The speed should match the speed specified in the software application. It does not need to be very fast. 9600 bps (bits per second or 'bauds') is adequate for most connections. I set my radio to 19200 bps. CAT TIME OUT TIMER sets the time that the radio will wait for a response to a CAT command. Leave it at the default setting unless you are experiencing CAT control problems. CAT RTS is not the PTT control line. That is set elsewhere and relates to the second COM port. It is used as an interrupt. When it is set to 'on' the radio monitors the RTS line on the CAT COM port and will respond to changes initiated by the PC software. When it is set to 'off,' the radio does not monitor the RTS line on the CAT COM port.

QMB CH: sets the number of quick memory slots associated with the QMB button. The default is 5 channels, but you can increase it to 10 channels. Each time a channel is saved it is placed into the top memory slot and all the previously saved frequencies move down one place. The one at the bottom of the list is discarded.

You can use this feature as a 'Scratch pad' to save a frequency that you want to return to. For instance, a station that you want to work but which is in a QSO with somebody else. Or you can use it to store frequencies that you use a lot, such as Net frequencies.

Turning **MEM GROUP** on splits the 99 stored memory slots into five groups. When they are split, you can scan a specific group, or select <FUNC> <MEM CH> to step through them using the MULTI control. Or by associating memory channel selection to the big VFO ring using the CS button.

TIP: Selecting memory channels will change the VFO frequency and mode if you have pressed the V/M button to select memory channel mode.

It is quite handy to save all the stored frequencies for a particular band into a group. That way the radio will not jump all over the HF band when you scan the memory slots. For example, you might put all the 6m repeater and simplex channels into group five. Or you might put all the international beacon frequencies into group two.

There are five general-purpose groups and two special groups. You cannot save a frequency to a memory position and then select which group it goes into. Instead, you must save the frequency into a memory slot that is in the wanted group. If you want to put a frequency into group 5 you have to store it in one of the slots between 80 and 99. You can't name, the groups either.

Memory channel	Group
1-19	1
20-39	2
40-59	3
60-69	4
80-99	5
Nine PMS scan edges	6 (PMS)
Ten 5 MHz (60m band) channels	5M

PMS = programmable memory scan. The eighteen memory slots hold the lower and upper frequency limits for nine scan ranges.

NOTE: You cannot use the GROUP menu item to change between groups unless the radio is in memory channel mode. In VFO mode you can only display the current group setting. This is mildly annoying and to my mind an unnecessary constraint.

If you have groups turned on, MEM CH can only select memory slots that are within the currently selected group. To select a memory channel in a different group, you have to select memory channel mode with V/M, then select the group, and then select the stored channel.

Split settings: The two split settings affect what happens when you press and hold the SPLIT button. I found the descriptions in the manual rather confusing, and it took a while to work it out.

QUICK SPLIT INPUT acts differently from what I expected. ON means quick split is <u>not</u> enabled and OFF means that quick split is enabled.

When QUICK SPLIT INPUT is on, and you press and hold the SPLIT button a popup box appears on the screen allowing you to select the split offset that you want. Touch 5 and then kHz if you want a five-kilohertz split. Touch -, 2, and then kHz if you want a negative two-kilohertz split.

When QUICK SPLIT INPUT is off, and you press and hold the SPLIT button, the sub-VFO will be set to transmit at the offset determined by the QUICK SPLIT FREQUENCY. In this mode, press and holding the SPLIT button a second time moves the transmitter further away from the receiver frequency by the same amount. i.e. if the offset is 5 kHz, the transmitter frequency will move up 5 kHz each time you press and hold the SPLIT button. I am not sure that is particularly useful since you can easily select the sub-VFO by touching the VFO display or pressing SUB and tune it with the VFO knob. I plan to leave this setting in the default OFF position.

WARNING. It is very easy to transmit on a different band or mode. If you just select SPLIT without press and holding the SPLIT button to set the split offset, the transceiver will happily transmit on whatever frequency and mode the sub-VFO is on.

The QUICK SPLIT FREQ setting is only relevant if QUICK SPLIT INPUT is set to off. In other words, if 'quick split' is enabled. You can set it anywhere from -20 kHz to +20 kHz. The default is 5 kHz which is good for upper sideband operation on SSB. You would probably set a 2 kHz offset for CW operation and a negative split for lower sideband.

TX TIME OUT TIMER: The timer limits the time that the transmitter will transmit continuously. You can set it to OFF or between one minute and 30 minutes. If you are prone to long overs, this might be a benefit. You get a beep warning when you are close to the set time, and the transceiver will return to receiving after the time has elapsed. Releasing the mic PTT every few minutes lets the 'other bloke' have a go and resets the timer.

It can be a good thing to use the timer if you are controlling the transmitter from a digital modes program or a connected device. Sometimes software or hardware glitches may cause an application to hang, and the transmitter will go into a permanent transmit mode. In the SSB Data mode, there probably won't be any modulating signal, but if it is a CW, RTTY, or FM signal the radio could be transmitting at full power.

Microphone scanning functions: The supplied SSM-75G hand microphone and many after-market microphones have handy buttons for controlling functions on the radio. The UP and DOWN buttons on the top of the hand microphone step the VFO up or down by 10 Hz. Or 5 Hz if you selected that option. The MIC SCAN setting affects the press and hold function of the UP and DOWN buttons. With the control set to ON, press and holding UP or DOWN starts a scan. Scanning will continue after you release the button until you press it again, press PTT, or touch the VFO. With the control set to OFF, press and holding UP or DOWN makes the VFO move up or down rapidly until you release the button.

MIC SCAN RESUME: affects what happens when the receiver encounters a signal during a scan that has been initiated by press and holding the UP or DOWN button on the microphone.

Mode	MIC SCAN RESUME	A signal is received or the squelch is open	No signal received and squelch closed
SSB or CW	PAUSE	Slower scan	Fast scan
SSB or CW	TIME	Slower scan	Fast scan

| FM and other modes | PAUSE | The scan stops until the signal disappears and the squelch closes | Fast scan |
| FM and other modes | TIME See note. | Scan waits for 5 seconds then starts to scan. If the squelch is still open it will pause again. If the squelch is closed it will resume scanning. | Fast scan |

Note: as far as I can tell you can't change the 5 second pause time.

REF FREQ FINE ADJ: The FTDX101 does not have an input for the connection of a reference frequency oscillator. It doesn't need one because the local oscillator is a 0.1 ppm high stability TCXO (temperature-controlled crystal oscillator). However, if you feel the need, you can calibrate the receiver frequency by netting to a standard frequency broadcast such as WWV or WWVH and adjusting the REF FREQ FINE ADJ control. Or you can measure the transmitter frequency using a frequency counter connected via a suitable coupler and RF power attenuators. The counter must be very stable and/or locked to a GPS referenced clock or a Rubidium frequency standard. Remember to apply any offset that might apply to the transceiver mode. Any adjustment made using FREQ FINE ADJ will affect both the receiver calibration and the transmitter frequency.

The TCXO in the radio is much more accurate than most frequency counters. I recommend that you do not make any changes to this setting unless you are 100% confident that fiddling with it is not going to make the accuracy of your transceiver worse.

CS DIAL: This control sets the function of the big VFO ring (MPVD) that will be activated when you press the CS (Custom Select) button. You can choose from the same list of options by press and holding the CS button. Make a selection here and when you press the CS button to the right side of the VFO knob, turning the VFO ring will adjust the chosen setting. I usually leave it set to the spectrum scope display LEVEL because I constantly have to adjust it.

The selections include.

Selection	Function	Note
RF POWER	Transmitter power	Might be useful if you are setting the level into a linear amplifier
MONI LVL	Monitor level	Normally a set and forget control
DNR LVL	Digital noise reduction	Could be useful if the band is noisy

NB LVL	Noise blanker level	May be useful if there is intermittent spikey noise
VOX GAIN	VOX gain	Normally a set and forget control
VOX DELAY	VOX delay	Normally a set and forget control
ANTI VOX	Anti VOX setting	May be useful if the background noise in your shack is variable
STEP DIAL	Fast tuning rate	Easy way to do 1 kHz tuning step, or you could just press FAST
MEM CH	Select a memory channel	Useful if you are in memory tune mode and the MULTI knob is assigned to a different function
GROUP	In memory tune mode it will select the group if groups are engaged	Has the same effect as assigning GROUP to the MULTI knob
R. FIL	Change the receiver roofing filter	Pointless. You can do this by touching R.FIL on the screen.
LEVEL	Adjust the level of the spectrum scope and waterfall display	Very useful for setting the level of the spectrum display.

TIP: *The big VFO ring (MPVD) is also be used to control the sub-VFO frequency <MAIN/SUB>, the VC-Tune <VC_TUNE>, or the Clarifier <CLAR>.*

KEYBOARD LANGUAGE:

There is a range of different keyboard layout options. The radio defaults to different options depending on the transceiver 'country' version. My radio defaults to ENGLISH (US). As far as I can tell this only affects the keyboard mapping of external keyboards plugged into a front panel USB port. The internal keyboard, displayed when setting up text macros, seems to be unaffected by this setting.

RX DSP

The RX DSP menu provides settings for the filters which are implemented in the IF DSP stage of each receiver.

APF width sets the bandwidth of the CW APF (audio peaking filter). Narrow, Medium, or Wide. A narrow setting will eliminate all other signals, but the signal needs to be precisely tuned. You can fine-tune the filter using the APF knob or tune the VFO using the CW tuning indicator underneath the DSP filter function spectrum display. The wider settings are easier to tune but you may hear more noise or interference from other stations operating very close to the frequency.

Contour level. The contour control provides a small dip or peak in the IF filter bandwidth. It is represented with a blue dip (or peak) on the DSP filter function spectrum display. You can use the CONT knob to set the dip (or peak) at any frequency across the IF bandwidth. The function can be used as a tone control to boost or reduce the treble or bass response, by setting the contour at the high or low end of the IF passband. Or it can be used as a 'mini notch filter' to reduce or eliminate weak interfering signals. The control is adjustable from a -40 dB dip to a +20 dB boost, with a default of -15 dB. *This range is not correctly stated in the current issue of the Yaesu manual.*

Contour width sets the width of the contour dip or peak. I can't see much point in changing from the Q=10 default setting. But you can adjust it from 1 to 11, "just like in the Spinal Tap movie." I guess a narrow setting might be useful if you are using the contour control to reduce low-level interference.

DNR level. The digital noise reduction can be set from 1 to 15 on each band. The default setting is 1. The DNR can affect the quality of the received signal. High levels will make the audio sound hollow and artificial, "like a Dalek." I suggest that you slowly increase the DNR level until you reach a level that reaches a compromise between achieving a good level of noise reduction and the wanted station not sounding totally awful.

TIP: You can assign DNR level to the MULTI knob for a few seconds by press and holding the DNR button. But you get more time to make adjustments if you press and hold the CS button and assign DNR LEVEL to the big VFO tuning ring.

IF notch width is used to change the IF notch from the default wide setting to a narrow setting. The wide setting is great for noisy or multi-carrier interference such as ADSL line noise. But the notch will attenuate more of the wanted signal. The narrow notch is better for single carrier "birdies," and it will have less effect on the quality of the wanted signal. I set the notch width to 'narrow' and only change it to 'wide' if the notch is not effectively eliminating the interfering signal.

TX AUDIO

You should definitely set **Proc Level** to COMP. Leaving it set to AMC means you cannot adjust the speech processor level. The control sets the function of the PROC/PITCH (outer) knob when you are in a voice mode.

Function of the PROC/PITCH (outer) knob		
Proc Level setting	Speech processor OFF	Speech processor ON
COMP	AMC Out Level	Speech PROC LEVEL
AMC	AMC Out Level	AMC Out Level

The function of the Proc Level control has led to much confusion in the online forums because its role has been changed as a result of an early firmware upgrade. Initially, you could turn off the AMC (automatic microphone level control) but now you can't.

The **AMC release time** sets the decay time after the AMC acts to reduce excessive microphone audio level. It is easier to adjust the AMC Out Level if the release time is set to FAST. The default level is MID. When operating, I can't tell the difference.

The **Parametric Equalizer** settings were discussed on page 18. You can tailor the audio frequency response of your transmitted signal using the three-band parametric equalizer. There are settings for when you have the speech compressor turned off and for when you have it turned on.

TX GENERAL

HF, 50M, 70M MAX POWER sets the maximum power of the transmitter on the relative bands.

Tip: Turning down the Max Power leaves a dead zone at the top of the RF POWER control. If you reduce the RF POWER control to the same output power level as the Max Power control and then reset Max Power to 100W (200W on the FTDX101MP) the transmitter power will stay low. However, if you don't adjust the RF POWER control and leave it set to its maximum when you reset Max Power to 100W (200W on the FTDX101MP) the transmitter power will return to full power.

AM MAX POWER does what it says. It is adjustable from 5 to 25 watts on the FTDX101D and from 5 to 50 watts on the FTDX101MP.

VOX Select is normally set to operate on input from the microphone (MIC), but you can set it to DATA so that it triggers on an audio signal applied to the USB cable or the RTTY/DATA jack. You could use the latter mode if you are completely unable to get the RTS/DTR signaling to work with your particular digital mode software. Or if you wanted to transmit a recording.

Emergency Freq TX (ON or OFF) allows the radio to transmit on the Alaska emergency frequency of 5167.5 kHz.

TUNING

This group sets the tuning step rate and the optical encoder setting for the main-VFO knob and the MPVD "big VFO ring."

- SSB/CW step size, 10 Hz or 5 Hz per step.
- RTTY/PSK step size, 10 Hz or 5 Hz per step.
- Step tuning rate for AM, FM, and all other modes
- VFO - Steps per revolution

Display settings

The Display Setting tab, <FUNC> <DISPLAY>, includes settings for the display, scope, and external monitor. These include display brightness, screen saver, mouse speed, the RBW, bandwidth, and sensitivity of the spectrum and waterfall display, and the resolution setting for an external monitor.

TIP: If you make any changes to the menu settings, it is a good idea to head to <FUNC> <EXTENSION SETTING> <MENU SAVE> <DONE> <NEW> <ENT> and save a backup to the SD card.

DISPLAY

My call lets you enter a 12 character personal message which is displayed on the 'opening screen' during the transceiver start-up. I had just enough room to enter ZL3DW FT101D. **My call time** sets the time that your message will be displayed. (OFF, 1, 2, 3, 4, or 5 seconds). Naturally, setting a longer time lengthens the time it takes for the transceiver to start. The actual time that the information is shown seems to be a little longer than the time you specify.

Screen Saver sets the time before the screen saver starts. The time resets if you touch the screen or make a change that affects the screen such as changing bands or pressing a button. The default is 60 minutes, but you can set the delay to Off, 15, 30, or 60 minutes. I don't recommend turning the screen saver off because you could eventually get a screen image burnt onto the display.

TFT contrast allows you to change the display contrast. High settings give a low contrast washed-out look. Low settings give a gloomy grey look but might be suitable if you like to operate in low-light situations. A setting of 6 to 8 would be good at night, but I leave my radio set to the default level of 10. **TFT Dimmer** adjusts the brightness of the display, from 0 to 20. It is a personal preference, but I like a setting of 12, a little less than the default level of 15.

LED Dimmer sets the brightness of the LEDs, from 0 to 20. This is a nice feature. You may want to dim the LEDs if you are working at night, especially if it's a contest and your eyes are tired. I set my radio for 6, a little lower than the default level of 10.

Mouse pointer speed does what it says. It adjusts the sensitivity of the mouse from 0 to 20 with a default setting of 20. I don't bother using a mouse, so I have not changed the setting.

Frequency style changes the weight of the font for the VFO numbers only. I rather like the elegant 'LIGHT' setting. The default setting is BOLD.

SCOPE

RBW (Resolution bandwidth): There are three choices of resolution bandwidth for the spectrum scope. I can't see any reason why you would want to change from the default HIGH setting. <FUNC> <DISPLAY SETTING> <SCOPE> <RBW>.

Scope Center: To change the scope center point, select <FUNC> <DISPLAY SETTING> <SCOPE> <SCOPE CTR> and select CAR POINT or FILTER.

If you choose FILTER, the receiver passband, which is usually be highlighted on the waterfall in a different shade or Color, will be placed right in the center of the screen, with the white center line in the middle of the receiver passband. If you turn on the Marker you will see the real VFO frequency on the left side of the receiver passband for USB, or the right side if the radio is set to LSB.

I strongly recommend using the 'carrier point' (CAR POINT) option which places the real VFO frequency on the white center line and displays the receiver passband on the right for USB, on the left for LSB, or in the center for AM, CW, or FM.

2D and 3DSS sensitivity: There are 'HI' and 'NORMAL' sensitivity settings for the 3DSS or waterfall and spectrum display. Try both settings and see which you prefer.

<FUNC> <DISPLAY SETTING> <SCOPE> <2D DISP SENSITIVITY>

<FUNC> <DISPLAY SETTING> <SCOPE> <3DSS DISP SENSITIVITY>

I prefer the HIGH setting for both the 3DSS display and the 2D display.

TIP: The 3DSS setting seems to make the display very slightly more sensitive when the HIGH setting is chosen. Which is what you would expect. However, the 2D setting seems to do the exact opposite. The display is much brighter on NORMAL than it is on HIGH.

EXT MONITOR

EXT DISPLAY: You must change a menu setting to use an external monitor.

<FUNC> <DISPLAY SETTING> <EXT MONITOR> <EXT DISPLAY> ON

PIXEL: I recommend you select the higher screen resolution option.

<FUNC> <DISPLAY SETTING> <EXT MONITOR> <PIXEL> 800x600

You get a big, low-resolution, image that looks the same as the touchscreen. But of course, it does not have touchscreen capability. It might be useful if you are demonstrating the radio at a public event or you have poor eyesight... but I don't see any appeal in trading the high-resolution touchscreen for a big blocky image on a monitor.

Extension settings

The Extension Setting menu, <FUNC> <EXTENSION SETTING>, sets the date and time, allows you to save and recall menu and memory contents from the SD card, list, update the current firmware, enter the calibration mode, and if all else fails, engage one of three levels of reset.

DATE & TIME

Setting the time and date. The radio has a clock, but it is never displayed on the screen. It is only used for timestamping saved files.

Select <FUNC> <EXTENSION SETTING> <DATE&TIME> and set the DAY, MONTH, YEAR, HOUR and MINUTE. If you want files date stamped with UTC date and time, enter UTC details. I believe it is easier to use local time for filenames.

SD CARD

The **SD Card** settings have already been discussed in the 'Using an SD card' chapter, on page 62 and the 'Firmware Update' chapter on page 63.

*TIP: If you insert the SD card with the transceiver turned on you are asked if you want to 'SETUP?' If you touch the YES option the **SD Card** setup menu will open so that you can format a new SD card, save or load the memory channel list, save or load a menu backup, look at the free space and capacity of the SD card, or carry out a firmware update.*

SOFTWARE VERSION

You can determine the currently installed firmware revisions using,

<FUNC> <EXTENSION SETTING> <SOFT VERSION>

Your firmware should be,

- Main CPU: (V01-14 or newer)
- Display CPU: (V01-06 or newer)
- Main DSP: (V01-05 or newer)
- SUB DSP: (V01-05 or newer)
- MAIN SDR (FPGA): (V02-03 or newer)
- SUB SDR (FPGA):   (V02-03 or newer)
- AF DSP   (V01-00 or newer)

CALIBRATION

The calibration section is for resetting the touchscreen calibration. You should only do this if a section of the touchscreen has stopped working correctly or part of the screen is not being displayed. For example, if the bottom set of Soft Key controls are partly hidden.

Touchscreen calibration is very easy to perform.

Select <FUNC> <EXTENSION SETTING> <CALIBRATION> <DONE>

A + sign will appear at the top left of the display. Touch the + symbol and it will reappear at the top right. Touch that and it will move to the bottom left, then the bottom right, and finally the center of the display. Touch each in turn and the display will be calibrated.

RESET

There are three levels of reset available. You should only do a reset as a last resort. Check the settings and connections first. Freeze-ups are usually fixed by turning off the radio and then restarting it. If the radio will not turn off, turn off or disconnect the DC supply. Or the Yaesu power supply unit in the case of the FTDX101MP. Then reconnect and restart the radio. If that fails, disconnect the CW key or paddle, all control cables, and the USB cable, leaving only the antenna cable(s) and the power supply cable connected. Turn off the transceiver. Turn off the DC supply or unplug the DC cable. After 30 seconds, reconnect and turn on the DC supply then turn on the radio. If you still have problems. Check the tightness of the DC cables where they are connected to the DC supply. Several readers have experienced problems with poorly crimped DC lugs.

OK here are the reset details!

Select <FUNC> <EXTENSION SETTING> <RESET> (reset option) <DONE>

Memory Clear resets all of the stored memory slots. Memory channel M-01 will be reset to 7.000 MHz LSB. The ten 5 MHz channels will be unaffected. You can restore the memory channels from a saved backup on the SD card. See the 'Using an SD card' chapter, on page 62.

Menu clear resets all menu settings to the factory defaults. This command does not affect the saved memory channels. You can restore the menu commands from a saved backup on the SD card. See the 'Using an SD card' chapter, on page 62.

All Reset clears the radio back to default settings including the menu setting and the memory channels. This reset will cause the radio to shut down and restart.

Front panel controls

This chapter describes the front panel controls. I have provided some additional details to supplement the information provided in the Yaesu manuals. The front panel controls are placed in functional groups making it easy to find what you want. There is a cluster of switches around the VFO knob, the band switches are grouped at the top, and the filtering controls are over to the right.

POWER

I guess you have already worked this one out. If you have entered your callsign it will be displayed under the Yaesu logo on the splash screen. (Page 9.)

Press and hold the POWER button to turn the radio on or off.

MAIN AF AND RF/SQL

The black inner knob controls the AF Gain (volume). The outer RF/SQL knob controls either the RF Gain or the Squelch.

➢ **RF gain and squelch**

The default setting for the RF/SQL control is for it to act as an RF Gain control. The RF Gain control provides a manual adjustment of the gain of the RF and I.F. stages. It can be very useful to turn down the RF gain to improve the signal to noise ratio of reasonably strong signals, especially on the noisy lower frequency bands. But most people leave it turned up to maximum all the time. In fact, the Yaesu manual states that the "RF/SQ knob is normally left in the fully clockwise position." I find it much more useful to change the menu setting so that the knob operates as a receiver squelch control.

If you operate mostly on 40m, 80m, and 160m my recommendation is to choose the RF gain option. If like me you favor the higher bands, set the menu to SQL.

The menu setting affects both receivers.
<FUNC> <OPERATION SETTING> <GENERAL> <RF/SQL VR> (RF or SQL).

There is no indication of the squelch level when SQL is selected, but a green BUSY indicator is displayed under the S meter when the squelch is open.

SUB-RECEIVER AF AND RF/SQL

The inner knob controls the AF Gain (volume). The outer RF/SQL knob controls either the RF Gain or the Squelch. If SQL has been selected, a green 'BUSY' indicator below the sub-receiver S meter indicates that the receiver squelch is open. The squelch or RF gain menu setting affects both receivers.

BAND BUTTONS

The band buttons are the top row of buttons above the VFO cluster. They let you select one of the twelve amateur radio bands that the transceiver covers. The 70/GEN button selects the 'general coverage' receiver. The radio will switch bands automatically if you tune the VFO to a frequency in an amateur radio band. A white LED indicates the current main-VFO band. Press SUB to change the sub-VFO band. The blue LED indicates the sub-VFO band.

Each band has Yaesu's standard 'band stacking register. Pressing a band button sets the radio to the last frequency and mode you used on the selected band. Pressing it again moves you to the frequency and mode you used before that, and pressing it again moves you to the frequency and mode you used before that. Each band button holds three frequencies.

TIP: You can manually program the band stacking register. Press a band button and tune the VFO to the frequency you want to save. Then press the same band button again to load the frequency into the band stacking register. If desired, you can move to a second frequency then press the band button to load that frequency into the band stacking register. Move to a third frequency then press the band button again to load the third frequency into the band stacking register. Not that these memory slots are volatile. They can be overwritten when you switch bands.

Unlike some radios, you cannot use the band button keys to directly enter a frequency. You can type in a frequency after touching the three Hz digits on the VFO frequency display.

Press and holding a band button turns on a small orange LED above the button. This has no function other than serving as a reminder of what bands are available using a certain antenna, or what bands you are using during a contest.

TUNE

Pressing TUNE activates the automatic antenna tuner. An orange LED on the button indicates that the tuner is active. The antenna tuner will not automatically initiate a 'tune' operation if the SWR is high. You have to manually start a 'tune' by press and holding the TUNE button.

Holding the TUNE button down forces a 'manual' tune operation. A carrier of about 10 Watts is generated and the LED flashes while the tuner is operating. You will hear relays clicking as the tuner attempts to find a match.

The tuner remembers the settings for each frequency that you transmit on, within 10 kHz band segments.

There is more about using the tuner on page 97 and page 78.

VOX

VOX can be activated when the radio is set to SSB, AM, FM, or DATA.

VOX stands for voice-operated switch. When VOX is on, the radio will transmit when you talk into the microphone without you having to press the PTT button.

Using VOX is popular if you are using a headset or a desk microphone. Touching the VOX key turns the VOX on or off. An amber indicator indicates VOX mode.

VOX settings are covered on page 79.

MOX

MOX stands for 'manually operated switch.' The switch is wired in parallel with the microphone PTT switch. It makes the transceiver switch to transmitting in the same way as pressing the mic switch. The radio will keep transmitting until you turn MOX off. A red LED on the button indicates that MOX is active.

Note that in CW mode MOX will allow you to send Morse code from a CW message, or the paddle or Morse key without having BK-IN turned on.

The microphone is live if you turn MOX on in the voice modes, so you will transmit any conversations, phone calls, talking to the cat, praising the dog, background music, or audio from another radio. The voice keyer will operate while the radio is being keyed via MOX or the microphone, but as usual, BK-IN must be turned on.

The FUNC button is inactive while the radio is transmitting. Although the screenshot mode still works.

ZIN/SPOT

ZIN is an auto-tune feature that pulls the receiver frequency onto a CW signal by matching the CW note with the Keyer Pitch. Netting it to the transmit frequency.

TIP: The manual reference on page 33 is misleading. It says to press the SELECT switch.

While receiving a CW signal, press ZIN/SPOT to pull the receiver VFO so that the received tone is the same as the keyer pitch. At that point, the transmitter will send on exactly the same frequency as you are receiving.

Most of the time it works very well, although you may have to press the button two or three times to get exactly on frequency. But it does not always work. Pressing the ZIN/SPOT switch can sometimes move you in steps away from the wanted frequency. The ZIN/SPOT button is only active in CW mode.

If you press and hold the ZIN/SPOT button the 'SPOT' keyer pitch tone will be heard, so you can compare the tone of the incoming signal with the keyer pitch. You can tune the VFO and 'zero beat' the audio from the two sources.

DISP

The DISP button is not active when the transceiver is in MONO mode, i.e. it only works if MONO is white.

Press DISP repeatedly to cycle through the four display options.

1. One big spectrum display. Can be 3DSS or normal.
2. Two spectrum displays SUB above MAIN. Either or both can be 3DSS or normal. The touchscreen controls affect the Main or sub receiver according to which receiver is selected by touching the VFO frequency display or by using the MAIN and SUB buttons.
3. Two spectrum displays, with MAIN on the left and SUB on the right. Either or both can be 3DSS or normal. The touchscreen controls affect the Main or sub receiver according to which receiver is selected by touching the VFO frequency display or by using the MAIN and SUB buttons.
4. The spectrum displays are replaced with the big meters and larger versions of the DSP filter function displays.

Modifiers

- The DISP button will not work if the display is in MONO mode.
- Pressing the DISP button deactivates the MULTI display.
- Pressing the DISP button deactivates the HOLD display.
- Touching EXPAND and then using the DISP button cycles, through slightly larger spectrum displays with the usual reduced metering. The last option with no spectrum displays is unaffected by the status of the EXPAND control.

S.MENU

The S.Menu button gives you fast access to the five spectrum display controls. You can also access them from the orange group in the FUNC menu. The functions are fully described on page 115.

FUNC

Pressing the FUNC button opens the Function menu which is described in the Function Menu chapter and the four following chapters, starting on page 115.

Press and holding the FUNC button takes a screenshot of the touchscreen display and saves it to the 'FTDX101\Capture' directory on the SD card.

MULTI

The MULTI knob is used to move through menu choices and memory channels. Pressing the MULTI button usually makes a selection from the current menu list.

The item currently allocated to the MULTI knob is displayed above the FUNC button. Other items may pop up for a few seconds. For example, press and holding an NB button pops up a level setting option that can be changed by turning the MULTI knob. After a few seconds, the popup disappears and the MULTI knob function returns to the item displayed above the FUNC button. You have to be quick to make the adjustment before the popup disappears.

MODE, SSB, AND CW BUTTONS

The **SSB button** selects SSB mode. Pressing it once sets the last used SSB mode on the current band. USB for bands above 10 MHz and LSB for bands below 10 MHz. Pressing it again selects the other sideband. There is no press and hold function.

The **CW button** selects CW mode. Pressing it once sets the last used CW mode on the current band. Either CW-U (CW on the upper sideband) or CW-L (CW on the lower sideband. Pick one and use it all the time. There is no press and hold function.

Pressing the **MODE button** selects the last mode that was selected using the MODE button. Having separate buttons for MODE, SSB, and CW is a real advantage. It means you can select your favorite mode at the press of a button.

Press and hold the MODE button to select any mode, LSB, USB, CW-L, CW-U, AM, AM-N, FM, FM-N, DATA-L, DATA-U. DATA-FM, DATA-FM-N, RTTY-L, RTTY-U, PSK, or PRESET. The modes are described in earlier chapters.

BK-IN

You must have BK-IN turned on to transmit CW. You can leave it on all the time. BK-IN is indicated with an orange LED in the button. The setting is remembered when you change modes. With BK-IN off you will hear the CW signal but it will not be transmitted unless the MOX button, the microphone PTT, or an external PTT signal perhaps from a footswitch, is activated.

There is no front panel control for selecting full or semi break-in. I guess most people use one or the other most of the time. Select SEMI or FULL Break-in using <FUNC> <CW SETTING> <MODE CW> <CW BK-IN TYPE>.

You must have BK-IN turned on to transmit the voice keyer messages.

You do not have to have BK-IN turned on to transmit RTTY or PSK messages.

MONI

MONI is the transmit monitor. It allows you to hear the signal that you are transmitting. If you are operating in SSB or AM mode, it is easier to hear and less distracting if you use headphones. You will hear the voice keyer audio, even if MONI is turned off. You must have the MONI turned on to hear the CW sidetone. The transmit monitor is not available in the FM modes.

The audio for the transmit monitor is taken from the transmitter I.F. after the modulator but before up-conversion to the final transmit frequency. This makes it especially useful for evaluating the quality and tone of the signal that you are transmitting. Use it when you adjust the parametric equalizer and speech processor (compressor) settings.

Press and hold the MONI button and use the MULTI knob to adjust the monitor level. The level is indicated on a popup, but it is better to just adjust it until you are happy with the volume. Or you can use <FUNC> <MONI LEVEL> to allocate the monitor level to the MULTI knob. Changing the monitor level on CW will not affect the monitor level when you are in SSB mode and vice versa.

THE MIC/SPEED AND PROC/PITCH CONTROLS

> Voice modes

Turning the inner MIC/SPEED knob changes the MIC GAIN

Pressing the MIC/SPEED knob engages the speech processor. This is indicated with a LED marked PROC.

When the speech processor is off, turning the outer PROC/PITCH knob adjusts the AMC OUT level. Setting the levels is covered under 'Getting ready for SSB operation' page 10.

When the speech processor is on, turning the outer knob adjusts the speech processor level, PROC LEVEL.

Check <FUNC> <OPERATION SETTING)> <TX AUDIO> <PROC LEVEL> <COMP>. *It must be set to COMP. If it is set to AMC, you cannot set the Speech Processor compression level.*

> CW mode

Turning the inner MIC/SPEED knob changes the CW keying speed.

Turning the outer PROC/PITCH knob changes the pitch of the CW signal.

Pressing the MIC/SPEED knob turns on the CW keyer, indicated with a LED marked KEYER. Leave it on all the time.

FAST

The FAST button located is to the left of the VFO, between the Main and Sub volume controls. It increases the VFO tuning rate from 10 Hz* to 100 Hz per step for the SSB, CW, and data modes, and from 5 kHz to 50 kHz per step for AM, FM, and FM data.

The FAST function is confirmed with an orange LED above the button.

If the MULTI knob has been set to control the STEP DIAL function. The FAST button will increase the rate from 2.5 kHz to 25 kHz per step. [The step rate is adjustable].

TIP: Press and holding FAST resets the three Hz digits to 000. If the scope is set to the FIX display mode, press and holding FAST resets the VFO to the beginning of the FIX spectrum range.

*The normal VFO rate can be set to 5 Hz or 10 Hz per step for SSB and CW, or RTTY and Data.

<FUNC> <OPERATION SETTING> <TUNING> <SSB/CW DIAL STEP>

<FUNC> <OPERATION SETTING> <TUNING> <RTTY/PSK DIAL STEP>

LOCK

The LOCK button is located between the Main and Sub volume controls, next to the FAST button. It locks the VFO tuning for the active VFO. If you press SUB or touch the sub-receiver VFO to change focus to the sub-VFO, you can lock the sub-VFO independently of the main-VFO. This will be great if you are operating as a 'Run' station in a contest, or you are a DX station working a pileup. You can lock your transmit VFO to make sure you don't accidentally move off your spot frequency, but you still have the freedom to tune your receiver across the band.

The LOCK function is confirmed with a LED above the button and 'Main Lock' or 'Sub Lock' popup indicator in the center of the screen.

SYNC

The SYNC button turns on the sub-receiver and locks the VFOs together so that turning the VFO knob affects the frequency of both the main channel and the sub receiver. But it does not net the frequencies together, so they might be on different bands or modes. The tuning rate is as per the main-VFO settings.

Press and hold the SYNC button to set the sub receiver to the same frequency and mode as the main receiver.

Note: If the sub-receiver is on and the main receiver is off, Sync will not turn on the main receiver.

SPLIT

The SPLIT button moves the transmitter to the sub-VFO frequency. But it does not turn on the sub receiver. You need to be very careful because the sub-VFO frequency might be on another band and/or set to a different mode.

Holding the SPLIT button until a double beep is heard can have two different effects depending on the QUICK SPLIT INPUT menu setting. Normally, with 'Quick Split' OFF, it sets the sub-VFO to the frequency and mode as the main-VFO plus or minus the shift set by the QUICK SPLIT FREQ menu. Press and holding the SPLIT button a second time moves the split offset further away from the receiver frequency by the same amount. For example, if the offset is 5 kHz, the transmitter frequency will move up 5 kHz each time you press and hold the SPLIT button. I am not sure that this is particularly useful since you can easily select the sub-VFO by touching the VFO display or pressing the SUB button and tune it with the VFO knob.

With QUICK SPLIT INPUT set to ON, holding down the SPLIT button brings up a window where you can set the split offset that you want to use. For example, touch 5 then kHz if you want a five-kilohertz split. Touch - 2 and then kHz if you want a negative two-kilohertz split.

<FUNC> <OPERATION SETTING> <GENERAL> <QUICK SPLIT INPUT>

<FUNC> <OPERATION SETTING> <GENERAL> <QUICK SPLIT FREQ>

For more information on how to use the Split mode effectively, see 'Operating Split' in the chapter on 'Operating the radio,' page 86.

CLAR (RX & TX)

The receiver clarifier is equivalent to RIT in earlier radios. It allows you to offset the receiver frequency while leaving the transmitter frequency the same as indicated on the VFO display. It is useful if you are chatting with a station that is slightly off-frequency. Rather than move the VFO which will cause the other station to adjust its frequency the next time you transmit, you can use the receiver clarifier to fine-tune the incoming signal. It is especially useful in a net where most stations are on frequency, but one is a little off frequency. The function is fully described back on page 71

The transmitter clarifier is equivalent to XIT in earlier radios. It allows you to offset the transmitter frequency from the receiver frequency. It can be another way of operating Split without using the split button. If the right-side TX button is selected and the sub-VFO is activated by touching the VFO frequency display or pressing the SUB button, the transmitter clarifier will move the frequency of the transmitter relative to the displayed sub-VFO frequency. The function is fully described back on page 81.

THE VFO CLUSTER

> RX, TX, Main⇔Sub, TX, RX

RX (left side) – turns on the main-VFO receiver. Indicated with a green LED.

TX (left side) – the radio will transmit on the frequency indicated on the main-VFO display. Indicated with a red LED.

MAIN⇔SUB – swaps the VFOs. Press and hold MAIN⇔SUB to set the sub-VFO to the same frequency and mode as the main-VFO.

RX (right side) – turns on the sub-VFO receiver. Indicated with a green LED.

TX (right side) – the radio will transmit on the frequency indicated on the sub-VFO display. Indicated with a red LED.

> VFO Knob

This is the main tuning knob. It's huge you can't miss it! You know what it does. The VFO control is very sensitive. I find myself bumping the radio off frequency quite a lot, especially when reaching around the VFO knob to adjust the outer tuning ring. It does not feel weighted and there is no apparent 'flywheel' action. i.e. the knob does not try to keep spinning when you stop turning the knob. There is a drag adjustment in a slot under the knob if you want to firm up the tuning, but I liked the VFO tuning action the way the radio was delivered. Unlike some other radios, there is no acceleration of the tuning rate if you turn the VFO quickly. In fact, the tuning rate seems to max out if you turn the knob very fast. The knob has a small dimple to aid fast tuning.

> MPVD (Multi-Purpose VFO Outer Dial) "The big VFO ring."

The VFO knob is surrounded by a cluster of buttons. Some of them are used to set the function performed by the outer tuning ring. Activating one of these buttons allocates the function to the big VFO tuning ring. They are CLAR which adjusts the receiver or transmitter clarifier offset, VC Tune, CS (custom select), and MAIN/SUB which allocates the sub-VFO tuning to the big VFO ring.

If none of these options is enabled, the big VFO ring tunes the main-VFO at the FAST rate. (Since firmware update V01-20).

> Fine Tuning

The FINE TUNING button is located at the lower left of the VFO button cluster. It selects a VFO tuning rate of 1 Hz per step for the SSB, CW, and data modes and 10 Hz per step for AM, FM, and FM data.

TIP: The FAST button has no effect if FINE TUNING has also been selected.

➤ QMB (Quick Memory Bank)

You can use this feature as a 'Scratch pad' to save a frequency that you want to return to. For instance, a station that you want to work but which is currently in a QSO with somebody else. Or you can use it to store frequencies that you use a lot, such as Net frequencies.

Press the QMB button to recall the last saved channel. Subsequent presses step through the five (or ten) saved frequencies. (Blank channels are not included).

Press and hold the QMB button to save the current VFO settings to the top of the stack. Each time a channel is saved it is placed into slot 1 and all the previously saved frequencies move down one slot. The one at the bottom of the list is discarded.

<FUNC> <OPERATION SETTING> <QMB CH> sets the number of quick memory slots associated with the QMB button. The default is 5 channels, but you can increase it to 10 channels.

➤ CLAR

The CLAR button in the VFO cluster allows you to change the RX and/or TX clarifier offset using the big VFO ring. Note that it does not actually apply the offset to the VFO unless the CLAR RX or CLAR TX button is turned on.

Turning on the CLAR RX or CLAR TX button automatically turns on the CLAR button. There is a more detailed explanation about the receiver clarifier on page 71 and the transmitter clarifier on page 81.

➤ VC TUNE

The VC TUNE button enables a tracking preselector. It is excellent if you have large interfering signals on the band such as you might experience during a contest weekend, or if nearby shortwave stations are creating intermodulation interference. You can manually adjust the VC Tune filter to help to eliminate interference, or just peak the received signal level shown on the S meter or spectrum display and let it automatically track any frequency changes that you make. If you switch bands the radio will remember whether the VC Tune was previously switched on for that band and the tuning setting. The beauty of the VC Tune system is that eliminates interference at the RF stage before the band pass filters and before it gets into the I.F. and the DSP stages of the receiver. As a result, it is effective at reducing intermodulation distortion.

Pressing the larger VC TUNE button allocates manual control of the VC Tune frequency to the big VFO ring. It also turns on the small VC TUNE button. A red bar graph above the DSP filter function display indicates the frequency of the VC Tune filter.

The VC tuner works at the received RF frequency right at the front-end of the receiver, not the audio signal shown on the filter display below the bar graph.

Press and holding either of the VC TUNE buttons resets the capacitor tuning so that the filter is in the center of the currently selected band. (Not necessarily the center of the bar graph). VC Tune does not work on the 5 MHz, 50 MHz (6m), or 70 MHz (4m) bands.

The FTDX101D has a VC Tune preselector unit on the main receiver. The FTDX101MP has a VC Tune preselector unit on both receivers.

> VC TUNE (mark II)

You may have noticed that there are two VC TUNE buttons in the VFO knob cluster. The smaller VC TUNE button turns on the VC Tune function without enabling the big VFO ring manual tuning.

> CS

CS stands for 'custom select.' It allows you to allocate one of twelve different functions to the big tuning ring. Since there is no control for adjusting the reference level of the spectrum scope, I usually allocate LEVEL to the tuning ring. Once allocated, pressing the CS button activates that function. If you use the clarifier, VC tune, or Main/Sub buttons, CS will turn off and the selected function will be allocated to the tuning ring. Pressing CS will reactivate the custom selection.

Press and hold CS to choose one of the 12 functions. RF Power, MONI level, DNR level, NB level, VOX gain, VOX delay, Anti-VOX, Step Dial, MEM CH, GROUP, R.FIL, or LEVEL. I cannot see any reason to set CS to R.FIL since it is always available by touching the screen. MEM CH is the same as allocating the memory channel to the MULTI knob. You must be in the 'memory channel' mode (V/M) for it to work.

> MAIN/SUB

The MAIN/SUB button on the lower right of the VFO knob allocates sub-VFO tuning to the big VFO ring. This is excellent when you are using the sub-receiver to tune through a pile-up of calling stations, or if you are tuning the sub-VFO on a different band to the main-VFO.

THE MEMORY BUTTONS

> V/M

V/M swaps between VFO and memory channel modes. Press and hold to display a list of the memory channels. Select a channel by touching the screen or by turning the MULTI knob. Then press (don't press and hold) the MULTI knob to set the channel into the VFO.

TIP: If you want to step through the memory channels you can use V/M to select the memory channel mode and then select <FUNC> <MEM CH> and turn the MULTI knob to step through the memory channels. Alternatively, you can use the big VFO ring to step through the memory channels if you allocate MEM CH to the CS control. Lastly, you can select the memory channel mode and then start a scan using <FUNC> <SCAN> or the Mic buttons.

> **M>V**

M>V is the memory channel to VFO button. It sets the contents of a memory slot into the VFO. It does not change the radio to the memory channel mode.

Press the M>V button to display the list of memory channels. Select a channel by touching the screen or turning the MULTI knob. Then press M>V again to load the data into the VFO. There is no press and hold function.

> **V<M**

V<M is the VFO to memory channel button. It is used to save frequencies to memory slots and recall memory channels while in the memory channel mode.

To **select a memory channel**, press the V<M button and select a channel by touching the screen or turning the MULTI knob. Then press (don't press and hold) the MULTI knob to set the channel into the memory channel display. If you move the VFO the radio will go into MT (memory tune) mode.

To **save a frequency and mode into a memory channel.** While in the VFO mode set the frequency and mode that you want to save. Then press the V<M button to enter memory channel mode and select a channel by touching the screen or by turning the MULTI knob. Then press and hold the V<M button to save the VFO frequency into the chosen memory slot.

THE DSP CLUSTER

The NB, DNR, and DNF buttons are grouped together because they form your toolkit for managing different types of interfering signals. See page 76 for a comparison between the NB, DNR, and DNF functions.

> **NB (Noise Blanker)**

Pressing the NB button or the sub-receiver NB button turns on the noise blanker for the relative receiver. Noise Blanker operation is indicated with an orange LED on the button. Holding down a NB button down brings up a sub-menu where you can adjust the noise blanker level, which is really the NB threshold. The noise blanker is designed to reduce or eliminate regular pulse-type noise such as car ignition noise. You may need to experiment with the level and the menu controls (NB width and NB Rejection) when tackling a particular noise problem. The full description of the noise blanker can be found on page 74.

➢ DNR (Digital Noise Reduction)

The noise reduction system in the FTDX101 is very effective. Pressing the main-receiver or sub-receiver DNR button turns the noise reduction on for the specified receiver. Digital noise reduction is indicated with an orange LED on the button. Holding the DNR button down brings up a sub-menu where you can adjust the noise reduction level from 1 to 15. Unlike the NB level, there are separate DNR settings for each receiver. See page 75.

➢ DNF

DNF is a DSP implemented automatic notch filter. It is effective at removing fixed carrier signals that are creating a beat note on SSB. It can remove the effect of multiple interfering carrier signals that are within the audio bandwidth. To test it, tune the radio so that you can hear a stray carrier signal. Looking at the audio filter response, turn on the DNF and you will see the filter remove the offending audio signal. You will still see the offending signal on the spectrum and waterfall display because that is showing the incoming RF signal, well before the digital notch filter in the DSP stage. You should not use DNF when receiving CW, PSK, or RTTY because it will filter out the wanted signal.

Tip: If DNR does not remove the interfering signal, try using the manual NOTCH instead. If the interference is wideband noise use DNR. For impulse noise use the noise blanker.

➢ NOTCH manual notch filter

Turning the NOTCH knob on the main or sub receiver automatically turns on the NOTCH filter button for that receiver. A red V indicating the notch appears on the DSP filter function display. Turning the knob changes the notch frequency and this is depicted by moving the slot across the audio spectrum and on a popup display.

I normally tune the control to the right spot by listening to the interfering signal disappear. Once you have adjusted the notch frequency you can toggle the filter on and off using the NOTCH button. Press and holding the NOTCH knob turns the filter off and resets it to the center of the filter passband.

You can change the width of the manual notch filter. The narrow setting will be fine for single carrier interference and it has less impact on the quality of the wanted signal. The default wide setting will be better at nulling out carriers that are modulated with noise or narrowband data.

<FUNC> <OPERATION SETTING> <RX DSP> <IF NOTCH WIDTH>

TIP: Yaesu recommends using the manual filter before trying the DNF notch filter if the interfering signal is very strong. This is because the manual notch is deeper than the auto notch.

➢ CONT

The contour control was new to me and I quite like it. It produces a small dip or boost in the audio frequency response which you can move across the audio passband from 50 to 3200 Hz. It's like a 'mini-notch.' You can hear the effect as you tune across the audio signal. It can be used as a sort of tone control, attenuating or boosting the low or high audio frequencies. Which could be useful if the received signal is "rumbly" or "hissy." It can also attenuate an interfering signal without notching out the audio frequency completely with the manual notch.

Turning the CONT knob (outside ring of the NOTCH knob) automatically turns on the CONT button. A blue 'dip' or 'bump' indicator is shown on the DSP filter function display. Once you have adjusted the contour frequency you can toggle the filter on and off using the CONT button. Press and holding the NOTCH knob turns the contour filter off and resets it to the center of the filter passband.

TIP: Contour adjustment of the higher audio frequencies has little effect on CW because the 600 Hz roofing filter attenuates those frequencies anyway. If you have a hearing impairment, for example, hearing loss at high frequencies you could set the contour level to a positive value and use the contour function to boost the high audio frequencies. There are no other tone controls so this might be the best solution.

You can change the bandwidth and the gain of the contour filter. You can even make it a hump so that it acts even more like a tone control, amplifying a part of the audio bandwidth. As an experiment, I set the control to +18 and could clearly see the increase in noise level as I tuned the contour control across the audio band.

<FUNC> <OPERATION SETTING> <RX DSP> <CONTOUR LEVEL> (-40 to +20)

<FUNC> <OPERATION SETTING> <RX DSP> <CONTOUR WIDTH> (1 to 11)

➢ APF

The APF (audio peak filter) button is for CW only. It is disabled on all other modes. Tune in a CW signal by ear or using ZIN or the audio bar-graph indicator and turn on APF. The very narrow audio filter eliminates nearly all the band noise revealing a clean CW signal. "It is fab!" The APF filter frequency is adjustable. Turning the CONT/APF knob while in the CW mode will activate the APF button and enable the knob. APF is adjustable over a range from -250 Hz to + 250 Hz. Press and hold NOTCH to reset the offset to zero. You can also change the bandwidth of the APF filter. The narrow filter will provide the cleanest sounding CW signal, but the signal has to be exactly on frequency. The wide setting is easier to tune, but you will hear more band noise.

Front panel connectors

USB JACKS

There are two USB 2.0 Type-A ports on the front panel under the Power button. They are for connecting a keyboard and/or a mouse.

➢ **USB mouse**

Connecting a computer mouse is easy. Just plug a USB mouse into one of the front panel USB ports and you are ready to go. The Logitech M310 and M305 wireless models will work, and probably some other mice. The right mouse button has no function and sadly the mouse wheel does nothing either. It could have been used to scroll the frequency… but no.

So, what can you do with the mouse? These items are not in the manual.

- The left mouse click has the same effect as touching the screen with your finger. You can change the meters, the DSP filter function display, the waterfall to spectrum ratio, the VFO frequencies, and operate the Soft Keys.
- If you are displaying both receivers, you can change the focus to either and adjust the relevant receiver by clicking on the spectrum display.
- If you press the FUNC button to open the FUNC menu, you can use the mouse to select any of the menu functions. Annoyingly you can't click the mouse on the Soft Key above the FUNC button, to open the FUNC menu, which would have made mouse operation much easier.

➢ **USB keyboard**

You can connect a USB keyboard to the radio and use it to fill in the TEXT messages. However, I can't see much point in plugging an external keyboard when there is an onscreen keyboard provided for this purpose. Unfortunately, you cannot use an external keyboard in conjunction with the digital mode decoders to send text directly from the keyboard.

TIP: Another odd feature is that if I plug in my Logitech wireless keyboard and mouse combo. The keyboard works, but the mouse doesn't.

KEY

The electronic keyer jack is for the connection of a CW bug, paddle, or straight key. It takes a standard ¼" (6.35 mm) stereo Phone plug. The wiring is standard. Dot on the Phone plug tip, Dash on the ring, and common on the sleeve. If you want to use a straight key use a stereo plug and wire the key to the tip and sleeve only.

Turn off the internal keyer by pushing the PROC/PITCH knob. See page 173 for a picture.

There is a second Key jack on the rear panel.

TIP: The 'key up' voltage on the front 'Key' jack is 3.3 volts, so it could be used with an Arduino or similar 3.3 volt logic device. The 'key up' voltage on the rear 'Key' jack is 5.0 volts.

PHONES

The PHONES jack is for connecting your stereo headphones. It takes a standard ¼" (6.35 mm) stereo Phone plug. The main receiver audio is on the left stereo channel and is connected to the plug tip. The sub receiver audio is on the right stereo channel and is connected to the plug ring.

The radio has only one internal speaker so there is no stereo effect using the internal speakers. There is a stereo AF OUT connector on the rear panel for connecting stereo PC-type amplified speakers, or for recording.

<FUNC> <OPERATION SETTING> <GENERAL> <Headphone Mix> is a useful setting. It changes the mix of main and sub-receiver audio signals into the headphones.

> SEPARATE – the main receiver is heard on the left channel and the sub-receiver is heard on the right channel.

> COMBINE-1 – this is a great option. It places the main receiver on the left channel but mixes in the sub-receiver at a lower level. On the right channel, you hear the sub-receiver with the main receiver at a lower level.

> COMBINE-2 – This is the mono option. The audio from both receivers is heard from both channels.

MIC JACK

The 8 pin Mic jack is for the microphone. It uses the standard Yaesu pinout, see page 32 in the Yaesu manual. There is no facility for the connection of a balanced microphone.

The SSM-75G microphone supplied with the radio has seven buttons in addition to the PTT switch. The UP and DWN buttons on the top of the microphone step the VFO frequency using the same step size as the VFO. The FAST button multiplies the step size by ten. In memory channel mode the UP and DWN buttons step through the memory channels. Press and hold to start a scan.

The MUTE button mutes the receiver audio. Very handy if someone comes into the shack while you are operating. But the button does not latch, you have to hold it down.

P1 is the same as pressing the MAIN-VFO button.

P2 is the same as pressing the SUB-VFO button.

P3 is the same as pressing the TX button on the main-VFO side.

P4 is the same as pressing the TX button on the sub-VFO side. (Handy for split operation).

SD CARD SLOT

The SD card is used to hold a backup of the radio's current menu settings and memory channels, or to save screenshots. It is also used for firmware updates.

You can use a 'full size' SD card, or a micro SD card in an adapter. Many micro SD cards are sold with a free SD adapter. The card can be a 2 Gb SD card or any SDHC card from 4 Gb up to 32 Gb. I am using a 16 Gb SanDisk SDHC card. The information screen <FUNC> <EXTENSION SETTING> <SD CARD> <INFORMATIONS> says that I have used 100 Mb so far, so buying a 16 Gb card like I did is probably overkill. I leave the SD card in the radio all the time.

TIP: don't pay extra for a fast 'Extreme' or 'React+' SDHC card. This is not an application where speed is important. Buy a slower / cheaper 'Ultra' or 'Select+' card. If you buy a micro SD card make sure you get one that comes with the plastic micro SD to full-size SD adapter.

Yaesu recommends formatting the SD card before using it for storage <FUNC> <EXTENSION SETTING> <SD CARD> <FORMAT>. I didn't bother and I have not experienced any problems.

The micro SD card slides into the adapter, contacts first, with its contacts facing down, the same side as the contacts on the adapter. The card will only fit one way.

Slide the full-size SD card (or card and adapter) into the SD card slot on the radio. Insert it contacts first, with the contacts facing down. The cut-off corner should be on the right. The card will only fit one way. It should slide easily and latch in with a click.

If the radio is turned on when you insert the SD card you will be presented with a popup menu that offers a 'Setup?' option (YES/NO). Selecting YES is perfectly safe. It just opens the <FUNC> <EXTENSION SETTING> <SD CARD> setup screen. Selecting NO closes the popup window.

To remove the SD card. Make sure that the transceiver is not writing data to the card, then push the card to un-latch it and remove your finger. The card should pop out far enough for you to be able to pull it out of the radio. It should slide easily. If it resists, repeat the 'push to unlatch' process.

Rear panel connectors

The only difference between the FTDX10D pictured below and the FTDX101MP is the DC power connector which is different because the FTDX101MP is powered from the supplied FPS-101 power supply unit which supplies 13.8 volts for the radio and 50 volts for the RF power amplifier.

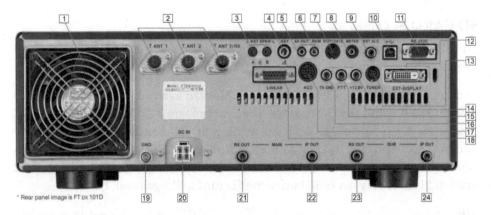

Figure 37: FTDX101D rear panel

Image from: https://silcom-ant.gr/product/yaesu-ftdx-101d/

1	Fan	13	Tuner (8 pin mini DIN)
2	ANT1, ANT2, ANT3/RX ANT	14	13.8 volts output (200 mA)
3	External speakers main & sub	15	PTT (input only)
4	Key	16	TX GND (PTT output to linear)
5	AF (audio) output	17	ACC
6	REM (remote)	18	Linear
7	RTTY/Data	19	Ground lug
8	Meter	20	DC input (FTDX101MP different)
9	Ext ALC (input)	21	RX OUT (main receiver)
10	USB Type B	22	IF OUT (main receiver)
11	RS-232 serial (DB-9)	23	RX OUT (sub receiver)
12	External Display (DVI-D)	24	IF OUT (sub receiver)

1. COOLING FAN

Make sure that the fan is not obstructed.

2. ANT1, ANT2, ANT3/RX

ANT 1, ANT 2, and ANT 3 are the primary antenna jacks that can be used for receiving and transmitting. You can select the antenna ports using the ANT Soft Key above the spectrum display. ANT3 can be configured to work as a receive-only antenna port using the antennas connected to ANT1 or ANT2 for transmitting, or you can disable transmitting altogether.

The connectors are standard 50 Ω SO-239 UHF jacks. They take a PL259 plug.

3. EXT SPEAKER

The two 'high level' mono speaker jacks on the rear of the radio are for connection to external speakers. If you want to use a set of amplified stereo PC speakers, use the lower output level AF-OUT stereo jack instead. The speaker jacks take mono 3.5mm phone connectors.

The effect of plugging in speakers depends on what you plug into, where.

A jack only: If you plug a speaker into the A speaker jack, it will replace the speaker in the radio. You will hear audio from both receivers (if they are both turned on) and the speaker in the radio will be off.

B jack only: If you plug a speaker into the B speaker jack, you will hear the main receiver on the external speaker and the sub receiver (if it is turned on) from the speaker in the radio.

A and B jacks: If you plug speakers into both speaker jacks, the main receiver will be heard from the speaker plugged into speaker jack B, and the sub receiver will be heard from the speaker plugged into speaker jack A. The speaker in the radio will be off.

4. KEY

The rear panel key jack is a ¼" (6.35 mm) Phone jack. It is for a straight key or a paddle. It has the same configuration as the front panel key jack except the 'key up' voltage is 5.0 volts. On the front panel Key jack, it is 3.3 volts. The rear panel Key jack will sink up to 3 mA.

5. AF-OUT

The AF-out jack takes a 3.5 mm stereo phone plug. It is ideal for connecting a pair of PC-type amplified speakers or for recording the receiver outputs. The main receiver audio is on the left stereo channel and is connected to the plug tip. The sub receiver audio is on the right stereo channel and is connected to the plug ring.

Main Sub Ground

6. REM

The REM mono mini phone jack is used to connect the FH-2 keypad. You can buy one at a reasonable price or construct your own with a few push buttons and resistors. I found this schematic on the Internet.

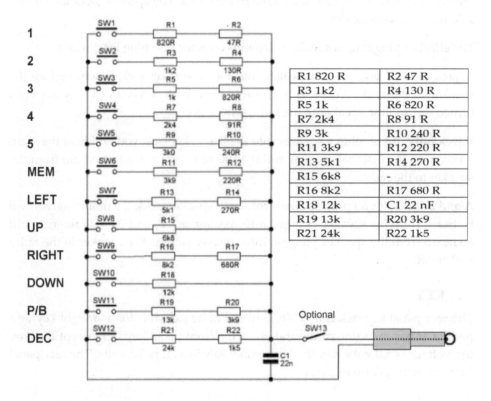

R1 820 R	R2 47 R
R3 1k2	R4 130 R
R5 1k	R6 820 R
R7 2k4	R8 91 R
R9 3k	R10 240 R
R11 3k9	R12 220 R
R13 5k1	R14 270 R
R15 6k8	-
R16 8k2	R17 680 R
R18 12k	C1 22 nF
R19 13k	R20 3k9
R21 24k	R22 1k5

Figure 38: Schematic of Yaesu style keyboard

7. RTTY/DATA

The RTTY/Data jack is for connection to a TNC for digital mode operation. These days you are more likely to use the USB cable unless you already have a TNC setup from a bygone era. It can be used for CW keying or HF Packet, but it is most commonly used for RTTY. There are pins for audio data input and output, PTT, ground, Squelch, and 'SHIFT' which is for FSK (or keying data) input.

TIP: The plug is a 6 pin mini-DIN connector. I have found the easiest way to source one is to chop the mouse off an old PS2 mouse and use the cable. If you don't have a PS2 mouse, op-shops often have a few because nobody buys them anymore.

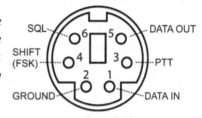

8. METER

The METER stereo mini-Phone jack is for the connection of external meters. On transmit, the meters display the same measurement as the meters on the radio.

The maximum output to the meters is around 3 volts, so if you are using ordinary mA panel meters you should add a voltage divider or a current limiting resistor in series with the meter. Using a 'trim-pot' will allow you to calibrate the meter.

The meter socket takes a stereo 3.5 mm phone plug. The main- receiver meter is connected to the connector tip and the sub-receiver meter is connected to the connector ring. The connector sleeve is common.

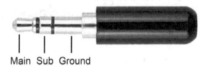

9. EXT ALC

The ALC and TX GND jacks are used to control a non-Yaesu linear amplifier. Use a stereo RCA audio cable to connect the TX GND (PTT) and ALC lines to your linear amp. The FTDX101 ALC line accepts a voltage between 0 and -4 Volts. Check that the amplifier output is compatible.

ALC should be configured so that it is not operating unless the transceiver is accidentally left at full power when driving the amplifier. Don't use it as a method of controlling the amplifier power. It should only be used as a failsafe in the event of a power setting mistake. ALC also helps to protect the linear amplifier from transmitting overshoot, although the transceiver may be too slow to react in time. Some transceivers emit a full power RF spike when you key the transmitter even when the RF power is set to a low level.

10. USB PORT

The USB connector is a USB 2.0 Type B port. You will need a Type A to Type B USB 2.0 cable to connect the radio to a PC.

The USB cable is used for CAT control of the transceiver and for transferring audio to and from a PC for external digital mode software. See the 'Setting up a USB connection' section on page 34.

The reason that the rear panel of the radio has a Type B USB port, and the front panel has Type-A USB ports, is that they have different functions. A USB Type-A port is a host port used for connecting a USB peripheral such as a keyboard, a mouse, or a USB memory stick. The USB Type B port is used when the device, in this case, the transceiver, is the peripheral connected to a remote host, in this case, your PC. USB Type-B connectors are used on items like scanners or printers.

11. RS-232

The RS-232 port can be used for CAT control and RTS/DTR signaling if you want to go "old-school" and if you can find a computer that has an RS-232 interface. The connector is a standard DB-9 serial interface. "Get into the nineties, man!" Use a USB cable.

12. EXT DISPLAY

The External Display port is a DVI-D connector for the connection of a computer monitor. You can buy a DVI-D to HDMI adapter if your display has HDMI, but no DVI-D input. The monitor must support 800x480 or 800x600 resolution. That could be a problem. My displays do not support 800x480. They will only go down to 800x600.

You must change a menu setting to use an external monitor.
<FUNC> <DISPLAY SETTING> <EXT MONITOR> <EXT DISPLAY> ON

And I recommend you select the higher screen resolution option.
<FUNC> <DISPLAY SETTING> <EXT MONITOR> <PIXEL> 800x600

You get a big, low-resolution, image that looks the same as the touchscreen. But of course, it does not have touchscreen capability. It might be useful if you are demonstrating the radio at a public event or you have poor eyesight… but I don't see any appeal in trading the high-resolution touchscreen for a big blocky image on a monitor.

13. TUNER

This 8 pin mini-DIN jack is for connecting a Yaesu FC-40 or Yaesu compatible antenna tuner.

14. +13.8V

The +13.8V RCA connector supplies 13.8 volts at up to 200 mA to supply a small external device such as a preamplifier or a TNC. You could also use it to power the light in a power meter.

15. PTT (INPUT ONLY)

This RCA connector is for PTT operation of the radio via a footswitch or similar. It has the same function as pressing the microphone PTT switch and is similar to operating the MOX switch. The open-circuit voltage is 5 volts and the circuit will draw 3 mA when activated.

16. TX GND (PTT OUTPUT TO LINEAR)

This RCA connector is the PTT output to a non-Yaesu linear amplifier. It is an open collector output capable of handling up to 60 volts DC at 200 mA, or 1 amp at 30 volts. That should be enough to switch the PTT of any linear amplifier.

17. ACC

The ACC (Accessory) jack is a 13 pin standard size DIN connector. It could be used for transceiver control from a variety of external hardware devices, including remote control of the radio. Although the manual does not suggest any options. The pinout is in the Yaesu manual.

18. LINEAR

The DB-15 LINEAR connector is used to provide band switching information, PTT, amplifier inhibit, and ALC information to the Yaesu VL-1000 linear amplifier. The band data can be used by other compatible amplifiers, or possibly for external tuner, antenna or preamplifier band switching. The pinout and band switching table is in the 'Using a linear amplifier' section on page 101.

19. GND

The radio ground connection should be connected to your shack ground in a 'star' rather than daisy chain format. The shack ground should be connected to an earth stake or earth mat outside. NOT to the mains earth. Earthing the radio can protect the radio from lightning static discharge (not direct lightning strikes). It can also improve noise performance.

20. DC IN

The FTDX101D requires a 13.8 volt (± 10%) regulated power supply capable of supplying at least 23 Amps (25A recommended). Use the supplied power lead. Never use a power lead without inline fuses.

BE VERY CAREFUL TO SUPPLY THE CORRECT DC POLARITY. Red is positive and black is negative.

The connector has a locking tab on the top. It may be covered by a black plastic connector protection. Squeeze the tab down before attempting to unplug the DC cable.

The FTDX101MP is supplied with the FPS-101 mains-powered DC supply, which supplies the radio with 13.8 volts and the final RF power amplifier with 50 volts. The supply voltage is 100-240V AC (50/60 Hz) at up to 9 Amps. The DC connector is different from the one fitted to the FTDX101D, so it is not possible to connect an incorrect power source.

21. RX OUT (MAIN RECEIVER)

RX OUT is a feed from the main receiver. It could be used for a connection to an SDR or another receiver. The take-off point is immediately before the 1st mixer. After the front-end attenuators and preamplifier, the VC Tune unit, and the bandpass filters.

22. I.F. OUT (MAIN RECEIVER)

The I.F. output is at 9 MHz. It is taken from a point after the 1st mixer but before the 9 MHz crystal filter, I.F. amplifier, and a low pass filter, so it may include harmonics and intermodulation products that will be filtered out by subsequent receiver stages. The most likely use of the I.F. output would be for connection to an external band scope or spectrum and waterfall display.

23. RX OUT (SUB RECEIVER)

RX OUT is a feed from the sub receiver. It could be used for a connection to an SDR or another receiver. The take-off point is immediately before the 1st mixer. After the front-end attenuators and preamplifier, the VC Tune unit, and the bandpass filters.

24. I.F. OUT (SUB RECEIVER)

The I.F. output is at 9 MHz. It is taken from a point after the 1st mixer but before the 9 MHz crystal filter, I.F. amplifier, and a low pass filter, so it may include harmonics and intermodulation products that will be filtered out by subsequent receiver stages. The most likely use of the I.F. output would be for connection to an external band scope or spectrum and waterfall display.

Troubleshooting

Most of the items covered in the troubleshooting chapter are not faults. However, they are conditions that might worry you if you encounter them while operating the radio. This chapter may help you if something unexpected happens. The items are not listed in any particular order, just the way I discovered them.

TRANSMIT POWER OVERSHOOT

Some early release models exhibited transmit power overshoot. This is a common problem with many transceiver brands and models. The issue arises when you use the RF POWER control to reduce the maximum power being applied to the input of a linear amplifier. The ALC in the radio is unable to react to the initial voice (or data audio) level and the radio transmits at full power for a few milliseconds before the ALC can react to reduce the RF output level. This short peak can trigger the protection circuits in the linear amplifier and trip out the power amp. The problem has been fixed now and radios produced after 18th June 2019 are unaffected. Yaesu has issued a warranty update for radios produced before that date.

If your radio's serial number is not on this list, the update has already been applied at the factory before shipping and you can ignore this notice. The update is not necessary if you do not intend to operate the FTDX101D with a linear amplifier.

Serial number range
9D010005 to 9D010005
9F010006 to 9F010023
9F020001 to 9F020038
9F020040 to 9F020047
9F020101 to 9F020221
9F020301 to 9F020329
9F020351 to 9F020382
9F030001 to 9F030073
9G020048
9G020223
9G020225 to 9G020229
9G020234 to 9G020234
9G020237 to 9G020237
9G020330 to 9G020349
9G020383 to 9G020395
9G030074 to 9G030094

YAESU SERIAL NUMBER IDENTIFICATION

It may be of interest to identify the date range of your Yaesu transceiver.

Digit	Function	Range
1	Last digit of the year of manufacture	9 = 2019 0 = 2020 1 = 2021 2 = 2022 etc.
2	Month of manufacture	C=Jan, D=Feb, E=Mar, F=Apr, G=May, H=Jun, I=Jul, J=Aug, K=Sep, L=Oct, M=Nov, N=Dec
3,4	Lot number	01 to 99
5,6,7,8	Serial number	0001 to 9999

For example, the first and last transceivers affected by the transmit power overshoot issue were.

9D010001   =   2019 Feb   Lot 01   Serial 0001
9G030094   =   2019 May   Lot 03   Serial 0094

RECEIVER TONE CONTROLS

There are no receiver tone controls or equalizers. But you can adjust the contour control to provide either a boost or a cut and position it at the low end of the audio spectrum to affect the bass response or the high end to affect the treble. See page 73.

BLACK SCREEN AND A YAESU LOGO ON THE SCREEN

Don't panic. The radio has a screen saver function to preserve your display just like your PC does. Touch the screen or press any button to restore normal operation.

You can turn it off, but I don't recommend doing that. The time before the screen saver starts is adjusted using <FUNC> <DISPLAY SETTING> <SCREEN SAVER>. You can choose from OFF, 15 minutes, 30 minutes, or 60 minutes.

WATERFALL TOO DARK, ONLY LARGE SPECTRUM PEAKS ARE SHOWING ON THE SPECTRUM DISPLAY.

This happens when the spectrum level is too low. Usually, if you change to a quieter band or select a different antenna. Select <FUNC> <LEVEL> and turn the MULTI knob clockwise to increase the spectrum level until the spectrum noise floor is just visible at the bottom of the window. The waterfall brightness should now be correct. You can also set the VFO ring (MPVD) to LEVEL by touch and holding the CS button and leaving the CS button turned on.

WATERFALL TOO LIGHT.

This happens when the spectrum level is too high. Usually, if you change to a noisier band or select a different antenna. Select <FUNC> <LEVEL> and turn the MULTI knob anti-clockwise to decrease the spectrum level until the spectrum noise floor is just visible at the bottom of the window. The waterfall brightness should now be correct. You can also set the VFO ring (MPVD) to LEVEL by touch and holding the CS button and leaving the CS button with its LED turned on.

PSK AND RTTY KEYER BUG

There was a very bad bug in the V01-14 firmware. It seems to have been fixed in the V01-20 firmware upgrade although nothing was stated in the update documentation.

When you have the decode window open and let's face it you do need to have the decode window open if you want to operate PSK or RTTY from the radio. I for one can't decode these modes in my head. The REC/PLAY message memories do not work properly. When you touch the message Soft Key or FH-2 key the radio does not transmit the message. It just sends the idle tone and will not stop. Sometimes if you press another key to stop the transmission the radio locks itself onto transmitting. All keys and buttons freeze and the radio has to be powered off to stop it from transmitting.

If you click the microphone PTT before touching the keyer number, it will usually, but not always send the macro. The macros seem to work fine if the decode window is not enabled. This means you can send PSK and RTTY, or you can decode PSK and RTTY but not at the same time.

GRUMBLES: It is very annoying that the RTTY or PSK message memory popup overlays the decode window, so you can't read the received messages. There are five blank Soft Keys on the decode window which could and should be used for sending the five message macros. The best answer is to buy or make an FH-2 keypad. Then you can send the messages without the popup window.

If you touch the display to clear the popup, you have to go all the way back, through pressing FUNC (you can't use the mouse) and then REC/PLAY to get it back again.

CW PADDLE NOT WORKING PROPERLY

This caused me some stress until I worked out what was wrong. It is not a fault, just a source of confusion. I had the menu set to where you can send CW while you are in the SSB mode turned on. I noticed in SSB that the keyer LED was on, so I pressed the MIC/SPEED button keyer to turn it off. After that, the CW paddle would not work, even when I switched back to CW mode. I messed around with the menu commands, but all I needed to do was turn on the internal keyer back on by pressing the MIC/SPEED (Keyer) knob.

VOICE MEMORY KEYER WON'T TRANSMIT

You must have BK-IN turned on to transmit the voice keyer messages. However, you can listen to your recordings by keying the messages with BK-IN turned off.

As of firmware revision V01-20A, the voice keyer will not transmit if PRESET is on. Even if you have changed from the DATA-U mode to SSB. The transmitter will turn on, but no power is generated. Make sure that PRESET is off. Press and hold MODE and make sure PRESET is grey. Turn BK-IN on, and then press <FUNC> <REC/PLAY> to show the voice message keyer.

RECORDED VOICE MESSAGES SOUND NOISY

This bug has been fixed in the V01-20 firmware release of April 2021. Earlier firmware versions have a problem that causes the audio quality to sound completely different when you are listening to the message with BK-IN turned off compared to when you are transmitting the message with BK-IN turned on. You hear the sub-receiver along with your recording. This happens even when the receiver is turned off and irrespective of the squelch control. The sub-receiver audio adds a lot of hiss and noise to the playback. Turn the volume of the sub-receiver down to zero and the message will sound the same as it does when you transmit it.

CAN'T TURN ON THE SPEECH PROCESSOR (COMPRESSOR)

If you have selected the mode where you can send CW while in SSB mode, you cannot use the MIC/SPEED (proc) knob to turn on the compressor. The workaround for this is to temporarily turn the CW on SSB function OFF using <FUNC> <CW SETTING> >MODE CW> <CW AUTO MODE> <OFF>, then turn the speech processor on by pressing the MIC/SPEED knob, then set the CW function back to ON or to 50M, then turn the keyer on by pressing the MIC/SPEED knob.

FT8 PRESET LOCKS UP TRANSCEIVER

The current firmware as I write this is V01-20A. Version 01-20 introduced the FT8 preset to the mode button. You are supposed to set up the preset parameters by touch and holding the PRESET icon on the MODE screen. On my radio, touch and holding the PRESET icon freezes up the transceiver. (See temporary solution)

Why didn't Yaesu add these presets to the RADIO SETTING menu?

> **Temporary solution**

Set <S.MENU> <PEAK> to LVL1, *I don't know why either, but it works*. Press <MODE> <touch and hold PRESET> and the setting menu should appear. If it does not appear, do not touch anything else except the blue PRESET icon. I think that this only has to be done the first time.

Change the pre-sets as required. Most of my ones were not set to the defaults which is odd. The Preset menu is discussed on page 33. If the radio freezes up at any stage during the process, turn the power off and restart the radio.

HIDDEN METERS

The audio spectrum scope can hide the meter displays. Touch the DSP filter function display to restore the meter display.

SPAN NOT FOLLOWING BAND SWITCH

A few times I have changed bands and the spectrum scope has not changed to the span setting last used on the new band. No cure.

LOST RECEIVER AUDIO

Turning on DNR and DNF at the same time caused a complete loss of audio from the main receiver. Required a power-off reset. Pressing any buttons in quick succession appears to confuse the operating system causing various lockups.

RECEIVER FAILING TO RECEIVE SIGNALS

I have had a situation where the receiver appeared to fail to receive signals. The S meter was at zero and the DSP filter function display showed an abnormally high noise level. I believe this was caused by accidentally bumping the AGC to OFF. Other possibilities are incorrect antenna selection, being in transmit mode, and wrong preamplifier or attenuator settings.

DSP FILTER FUNCTION DISPLAY NOT SHOWING SPECTRUM

Touch and holding the DSP filter function display turns off the spectrum display, leaving only the passband indicator and the notch or contour indicators if they are turned on. Touch and holding the DSP filter function display again should restore the spectrum display, but I have had it lock in the off position. Turning off the transceiver cured the problem.

DIGITAL MODE PROBLEMS

> **Digital mode audio signals**

The audio signal is sent over the USB cable to the PC in any mode. So, you can use your digital mode PC software to see and decode digital mode signals like RTTY, PSK, or FT8. But you cannot transmit audio digital mode signals unless you are in a data mode, usually DATA-U. Use DATA-FM for FM Packet radio on 6m. CW is digitally keyed, so it will work when the transceiver is in CW mode. FSK RTTY is also digitally keyed, so you should use the RTTY-L mode for that. Note that not all digital mode software supports FSK keying of the FTDX101.

> **PC digital mode software working on the wrong band.**

PC digital mode software uses CAT commands to read or set the active VFO frequency. It will probably display the frequency of the currently selected VFO. However, the audio signal displayed on the PC screen always comes from the main receiver on the left stereo channel. This can cause the curious case of the PC digital mode software displaying the sub-VFO frequency while the waterfall is displaying the audio from the main-VFO. It is best if you always use the main-VFO with PC digital mode software and make sure that you set the software for input from the left audio channel only.

The right stereo channel carries audio from the sub receiver. If either receiver is turned off with the RX buttons, the audio output will stop. The front panel volume controls do not affect the audio output to the PC over the USB cable, however, the squelch controls are active.

> **Phantom signals displayed on PC digital mode waterfall**

The FTDX101 outputs audio from the main receiver on the left stereo channel and audio from the sub receiver on the right stereo channel. If you have neglected to change the Windows mixer to a two-channel input and you turn on the sub receiver, you will get a mix of the audio from both receivers on the PC software waterfall. See 'Setting up a USB connection – audio codec' on page 36. The same problem will occur if you have set the PC digital mode software audio input for 'both' left and right audio channels. Unless you specifically want the audio from the sub receiver, make sure that you set the software for an input from the left channel only.

S METER STUCK AT A HIGH LEVEL

The S meter being fixed at a high level and not moving is an indication that the RF level has been turned down. Turn up the RF gain to the maximum or change the control to squelch instead of RF gain.

<FUNC> <OPERATION SETTING> <GENERAL> <RF/SQL VR>

The other possibility is that the radio is in transmit mode. Check the red TX icon below the meter is not on. TX is also indicated by a red LED on the band selector.

TRANSCEIVER STUCK IN TRANSMIT MODE

> **Transmitter held on by the USB COM port**

Or starting your PC digital modes software causes the transceiver to go into transmit mode.

This is not a fault, but it can be rather disconcerting. In SSB mode it is unlikely to damage anything as there will be very little power transmitted. However, in CW AM or FM mode, it will cause full power to be transmitted.

The problem is usually due to the RTS / DTS settings on the COM port being inappropriately controlled by the PC software. Normally the transceiver is set so that the RTS signal is PTT (turns transmit on) and the DTS signal is used to send CW or RTTY data. If the digital mode software has either RTS or DTR held 'active' it can make the transceiver switch to transmit. If this happens, check the COM port settings in the digital mode software and make sure that the RTS and DTR lines are set the same way as the transceiver, or to Always Off. This condition can be proved out of the radio by removing the USB cable.

It is usually possible to set the digital mode software to use a CAT command for PTT rather than the RTS (or DTR) line. But use RTS/DTR signaling if you can.

➤ **Transmitter held on in PSK or RTTY mode**

Firmware version V01-14 has a **bug in the message keyers** that causes the transceiver to go into transmit until manually stopped. See 'PSK and RTTY keyer bug' above. Usually, you can just touch the message Soft Key and the transmitter will stop, but sometimes touching the Soft Keys causes a system crash and the transmitter cannot be stopped. All the front panel buttons and Soft Keys freeze. The only way to stop the transmitter is to turn the power off using the normal power switch. When the radio re-starts it will be back to normal. This was fixed in V01-20. Upgrade to the latest firmware revision.

➤ **Other reasons the transmitter may be stuck on.**

Check that the front panel MOX button has not been pressed. It could also be the Morse key connection or the PTT. Check that the Microphone PTT button is not stuck on transmit and remove the cables from the rear panel PTT jack, and both KEY jacks. The ACC jack also has a PTT capability, which can turn on the transmitter. A stuck key on the external FH-2 keypad is also a possibility.

Basically, unplug all of the external cables except the antennas and the DC supply until the problem is resolved. Then problem-solve the cable connection that is causing the issue.

Turn off the power using the normal power switch. When the radio re-starts it will hopefully be back to normal.

TRANSCEIVER LOCKS UP

Unfortunately, it is very easy to crash the firmware in the radio causing the radio to freeze. Sometimes while transmitting. Generally, this seems to happen if you press too many buttons in close succession. It can also happen if the RTS and DTR settings are incorrectly set in an external digital modes program.

Normally you can turn the radio off with the ON/OFF power switch, but I have experienced freeze-ups when even that would not work, and I had to turn off the external 13.8 volt DC supply.

Strangely when I restored the volts, the radio came back up already turned on!

Changing the settings on the new PRESET function is also likely to cause the transceiver to crash! "Happy days."

MEMORY CHANNELS WILL NOT SELECT

There is some general weirdness about the memory channel selection if 'groups' is turned on. <FUNC> <OPERATION SETTING> <GENERAL> <MEM GROUP>

Constraint number 1: You cannot use the <FUNC> <GROUP> menu item to change between groups unless the radio is in memory channel mode (V/M). In VFO mode you can only display the current group setting. This is mildly annoying and to my mind an unnecessary constraint.

Constraint number 2: If you have groups turned on, you can only use <FUNC> <MEM CH> to select memory slots within the currently selected group. In other words, you have to select memory tune mode with V/M, then select the group with <FUNC> <GROUP>, and then you can use the <FUNC> <MEM CH> menu option to select a stored channel using the MULTI knob.

While having groups is a great idea, the difficulty of choosing memory channels makes using them a challenge. I am not sure that it is worth the hassle.

SWR METER DISAPPEARS WHEN I TRANSMIT

The second meter display can disappear when you transmit. This can happen if you are operating with split frequencies and you only have the main-VFO visible, i.e. MONO is blue. The display is normal while the radio is receiving. You can see two meters and the DSP filter function display. When you transmit in split mode, the VFO changes to the sub-VFO. You will see that the spectrum scope change from MAIN to SUB and the VFO frequency changes to the transmit frequency. If you happen to have set the sub-VFO to a single meter with a larger DSP filter function display you will lose the second meter (SWR) while you are transmitting. The spectrum display(s) also change to CENTER mode while you transmit.

SWR METER APPEARS WHEN I TRANSMIT.

This is the same effect as above. But in this case, it is quite a neat feature. Because you do not use the second meter while receiving, there is no point in having it on the screen. If you are operating with split frequencies and you only have the main-VFO visible, i.e., MONO is blue. You can set the main-VFO to show one meter and the larger DSP filter function display and the sub-VFO to show two meters and a small DSP filter function display. When you transmit the second meter will pop up to show you the SWR or ALC.

CW MODE - SIDETONE BUT NO TRANSMITTER POWER

This one has caught me out several times. CW will not be transmitted unless break-in has been turned on by pressing the BK-IN button.

Alternatively, with BK-IN turned off, keying the PTT line by pressing MOX, pressing the PTT button on the microphone, sending a CAT command from PC software, or switching the PTT line using the ACC connector, PTT jack, or LINEAR port will put the transmitter into transmit mode and the CW signal will be transmitted.

CW MODE – NO SIDETONE

You have to have MONI turned on to hear the sidetone. I had a situation where there was no sidetone when I used the CW paddle and MONI was on. Turning BK-IN on or off had no effect. There was what sounded like receiver noise, but the volume controls had no effect on either receiver, so I believe that it was digital noise not received spectrum noise. I turned off both receivers using the green RX buttons then turned the main receiver back on and the problem was resolved.

RF POWER METER NOT ACCURATE

There has been a report, 1 July 2020, that the RF power meter is not accurate when you turn down the power below 50 watts. I have not tested this, and I do not know if the problem has been fixed in subsequent firmware updates. I don't think that it is a very serious problem.

ANTENNA TUNER DOES NOT WORK ON 70 MHZ BAND

No. The tuner works from 1.8 MHz to 29.7 MHz. I believe that the only place this is mentioned in the Yaesu manual is on the specification page.

CALIBRATING THE SPECTRUM SCOPE FOR TRANSMITTING

I expect that few people will bother calibrating the spectrum scope for transmitting because there is no separate LEVEL adjustment ☹. Select the CW mode with the keyer turned off and BK-IN turned on. Key the transmitter with your Morse key and use the LEVEL control to set the top of the CW signal to the top of the spectrum display. That calibrates the top line of the spectrum scope to +50 dBm (100W) PEP or +53 dBm (200W) on the MP. The spectrum scope is factory calibrated to 5 dB per division. Now return to SSB. When you talk, the IMD (intermodulation distortion) products should not exceed +20 dBm [+23 dBm on the MP]. (30 dB below the PEP level). Put simply, the IMD products should not exceed the bottom four lines of the calibrated display. +23 dBm is 0.2 Watts. +20 dBm is 0.1 Watts.

Glossary

Term	Description
3DSS	The 'three dimensions signal stream' is a spectrum display that includes signal history similar to the information on a waterfall display. It is represented as a 3D image tapering back to give an illusion of depth.
6m, 10m 20m, 40m, 80m	50 MHz, 28 MHz, 14 MHz, 7 MHz, and 3.5 MHz amateur radio bands
59	Standard (default) signal report for amateur radio voice conversations. A report of '59' means excellent readability and strength.
599, 5NN	Standard (default) signal report for amateur radio CW conversations. A 599 report means perfect readability, strength, and tone. The 599 signal report is often used for digital modes as well. The 5NN version is faster to send using CW. It is often used as a signal report when working contest stations.
73	Morse code abbreviation 'best wishes, see you later.' It is used when you have finished transmitting at the end of the conversation.
.dll	Dynamic Link Library. A reusable software block which can be called from other programs.
A/D	Analog to digital
ACC	13 pin DIN accessory jack
ADC	Analog to digital converter or analog to digital conversion
AF	Audio frequency - nominally 20 to 20,000 Hz.
AFSK	Audio Frequency Shift Keying. An RTTY mode that uses tones rather than a digital signal to drive the SSB transmitter.
AGC	Automatic Gain Control. In the FTDX101 AGC is performed in the DSP stage at I.F or audio.
ALC	Automatic Level Control. There are two kinds used by the radio. One is the AMC (automatic microphone gain) used to ensure that the radio is not overmodulated. This is metered by the ALC meter. The other is the ALC control sent from a linear amplifier to the EXT ALC jack which ensures that it is never overdriven by the transceiver.
Algorithm	A process, or set of rules, to be followed in calculations or other problem-solving operations, especially by a computer. In DSP it is a mathematical formula, code block, or process that acts on the data signal stream to perform a particular function, for example, a noise filter.

AM	Amplitude modulation, (double sideband with carrier)
ANT	Abbreviation for antenna
APF	Audio Peak Filter. This is a very sharp filter that operates on the DSP audio signal. It is only available in the CW mode. The APF frequency can be offset using the CONT/APF knob.
ATT	Abbreviation for attenuator. The FTDX101 has front-end attenuator selections for 6 dB, 12 dB, or 18 dB.
Band scope	A band scope is a spectrum display of the frequencies above and below the frequency that the radio is tuned to. The center of the display is generally the frequency that you are listening to. This is different from a spectrum and waterfall display where you can listen to any frequency across the display.
Bit	Binary value 0 or 1.
BK_IN or BKIN	CW 'Break-in' the practice of receiving Morse code between the Morse characters or words that you are sending. The radio will not automatically transmit in the CW mode unless BK-IN is turned on.
BPSK	Binary phase-shift keying. Digital transmission mode using a 180-degree phase change to indicate the transition from a binary one to a binary zero. The PSK decoder in the FTDX101 supports BPSK31 and QPSK.
BW	Bandwidth. The range between two frequencies. For example, an audio passband from 200 Hz to 2800 Hz has a 2.6 kHz bandwidth.
Carrier	Usually refers to the transmission of an unmodulated RF signal. It is called a carrier because the modulation process modifies the unmodulated RF signal to carry the modulation information.
CAT	Computer-aided transceiver. Text strings are used to control a ham radio transceiver from a computer program.
CENTER	Sets the spectrum and waterfall display so that the main-VFO frequency is in the center of the display with a SPAN of frequencies on either side.
CODEC	Coder/decoder - a device or software used for encoding or decoding a digital data stream.
COM	Serial Communications port. In the FTDX101 the two Com Ports are 'virtual' serial ports for the data carried over the USB cable. There is also a standard RS-232 com port on the rear panel.
COMP	Speech compressor (processor). Increases the average power of your transmission by decreasing the dynamic range of the audio signal. i.e. it makes the quiet parts louder.
CPU	Central processing unit - usually a microprocessor. Can be implemented within an FPGA.

CQ	"Seek You" an abbreviation used by amateur radio operators when making a general call which anyone can answer.
CW	Continuous Wave. The mode used to send Morse Code.
D/A	Digital to analog.
DAC	Digital to analog converter or digital to analog conversion
data	A stream of binary digital bits carrying information
DATA	One of the data modes (RTTY-L, DATA-FM, or DATA-U) used to interface the radio with a PC digital mode program. You must be in a DATA mode to transmit from a PC digital mode program.
dB, dBm, dBc, dBV	The Decibel (dB) is a way of representing numbers using a logarithmic scale. Decibels are used to describe a ratio, i.e. the difference between two levels or numbers. They are often referenced to a fixed value such as a Volt (dBV), a milliwatt (dBm), or the carrier level (dBc). Decibels are also used to represent logarithmic units of gain or loss. An amplifier might have 3 dB of gain. An attenuator might have a 10 dB loss.
DC	Direct Current. You need a 25 Amp regulated 13.8V DC supply to power the radio.
Digital modes	Amateur radio transmission of digital information rather than voice. It can be text or data such as video, still pictures, or computer files, (PSK, RTTY, FT8, Olivia, SSTV, etc.)
DNF	The Digital Notch Filter automatically eliminates the effect of long-term interference signals such as carrier signals that are close to the wanted receiving frequency. Not effective against impulse noise.
DSP	Digital signal processing. The FTDX101 has Texas TMS320C6746 DSP integrated circuits for each receiver. DSP uses mathematical algorithms in computer firmware to manipulate digital signals in ways that are equivalent to functions performed on analog signals by hardware mixers, oscillators, filters, amplifiers, attenuators, modulators, and demodulators.
DTR	'Device terminal Ready' a com port control line often used for sending CW or FSK RTTY data over the CAT interface between the radio and a PC.
DVI-D	Digital Visual Interface is the video standard and connector used for the external display connection.
DX	Long-distance, or rare, or wanted by you, amateur radio station. The abbreviation comes from the Morse telegraphy code for 'distant exchange.'
DXCC	The DX Century Club. An awards program based around confirming contacts with 100 DXCC 'countries' or 'entities' on various modes and bands. The DXCC list of 304 currently acceptable DXCC entities is used as the worldwide standard for

	what is a separate country or a recognizably separate island or geographic region.
DXpedition	A DXpedition is a single, or group of, amateur radio operators who travel to a rare or difficult to contact location, for the purpose of making contacts with as many amateur radio operators as possible worldwide. They often activate rare DXCC entities or islands, and they may operate stations on several bands and modes simultaneously.
EXPD	Abbreviation for Expanded
EXT	Abbreviation for External
FFT	Fast Fourier Transformation – conversion of signals from the time domain to the frequency domain (and back).
FIX	Fixed mode. Sets the spectrum and waterfall display so that it displays the range frequencies between two pre-set frequencies determined a start frequency and the Span setting.
FM	Frequency Modulation. Used for repeaters on the 10m and 6m bands.
FPGA	Field Programmable Gate Array – a chip that can be programmed to act like logic circuits, memory, or a CPU.
FSK	Frequency Shift Keying.
FSK RTTY	FSK RTTY is keyed using a digital signal to offset the transmit frequency rather than AFSK which generates audio tones at the Mark and Space offsets from the VFO frequency.
GND GROUND	Ground. The earthing terminal for the radio. This should be connected to a 'telecommunications' ground spike, not the mains earth.
GPS	Global Positioning System. A network of satellites used for navigation, location, and very accurate time signals.
GPSDO	GPS disciplined oscillator. An oscillator locked to time signals received from GPS satellites
Hex	Hexadecimal – a base 16 number system used as a convenient way to represent binary numbers. For example, 1001 1000 in binary is equal to 98h or 152 in decimal.
HF	High Frequency (3 MHz -30 MHz)
Hz	Hertz is a unit of frequency. 1 Hz = 1 cycle per second.
IF or I.F.	Intermediate frequency = the Signal – LO (or Signal + LO) output of a mixer
IMD	Intermodulation distortion - interference/distortion caused by non-linear devices like mixers. There are IMD tests for receivers and transmitters. IMD performance of linear amplifiers can also be tested.

IPO	Intercept Point Optimization. Maximizes the dynamic range and enhances the close multi-signal and intermodulation characteristics of the receiver.
IQ	Refers to the I and Q data streams treated as a pair of signals. For example, a digital signal carrying both the I (incident) and Q (quadrature) data.
Key	A straight key, paddle, or bug, used to send Morse Code
kHz	Kilohertz is a unit of frequency. 1 kHz = 1 thousand cycles per second.
LAN	Local Area Network. The Ethernet and WIFI connected devices connected to an ADSL or fiber router at your house is a LAN.
LED	Light Emitting Diode
LSB	Lower sideband SSB transmission
m, 2m, 6m	Meter (US) or Metre EU). Often used to denote an amateur radio or shortwave band; e.g. 2m, 6m, 30m, 10m, where it denotes the approximate free-space wavelength of the radio frequency. Wavelength = 300 / frequency in MHz. A frequency range of 3 to 30 MHz has a corresponding wavelength of 100m to 10m.
Marker	The receive and transmit frequency markers indicate the frequency of the main-VFO (in FIX mode), the sub-VFO, and the Transmit frequency when using split.
MDS	Minimum discernible signal. A measurement of receiver sensitivity
MHz	Megahertz – unit of frequency = 1 million cycles per second.
MIC	Microphone
MT	Memory Tune. If you select a memory channel and then turn the VFO knob to a different frequency the memory number above the mode indicator will change from the memory number to MT. The radio will behave as if it is in VFO mode.
NB	Noise Blanker. A filter used to eliminate impulse noise
Net	Usually refers to an on-air meeting of a group of amateur operators. Also, to 'Net' to the CW receive frequency means to adjust the transmitter frequency or the receiver frequency so that you will transmit CW on exactly the same frequency as you are receiving.
NR	Noise Reduction. A filter used to eliminate continuous background noise
Onboard	A feature performed within the radio. Especially one that usually requires external software. For example, the radio has 'onboard' RTTY and PSK decoders.
PC	Personal Computer. For the examples throughout this book, it means a computer running Windows 10.

Pileup	A pileup is a situation when a large number of stations are trying to work a single station. For example, a DXpedition or a rare DXCC entity. Split operation is often employed to spread the pileup of calling stations over a range of frequencies.
Po	RF power output (meter)
PSK	Phase shift keying. Digital transmission mode using phase change to indicate the transition from a binary one to a binary zero. The PSK decoder in the FTDX101 supports three PSK modes; BPSK31, BPSK63, and QPSK31.
PMS	Programmable memory scan. Scans between frequencies stored in memory slots M-P1L and M-P1U through M-P9L and M-P9U.
PTT	Press to talk - the transmit button on a microphone – The PTT signal sets the radio and software to transmit mode.
QMB	Quick memory button. A short-term memory function, storing either five or ten frequencies along with the mode and filter settings.
QPSK	Quadrature phase-shift keying. Digital transmission mode using 90-degree phase changes to indicate four two-bit binary states 00,01,10,11. The QPSK decoder in the FTDX101 supports QPSK31. QPSK is faster than BPSK31 but uses more bandwidth and is not as easy to decode. It usually has built-in error correction.
QRP	Q code - low power operation (usually less than 10 Watts).
QSO	Q code – an amateur radio conversation or "contact".
QSK	Q code – fast transmit to receive switching which allows Morse code to be received in the gaps between the CW characters that you are sending.
QSY	Q code – a request or decision to change to another frequency.
RBW	Resolution Bandwidth is the ability of the spectrum scope (spectrum and waterfall) to distinguish between signals that are on frequencies that are very close together. A high RBW can display signals that are closer together but requires more processing power and speed.
RF	Radio Frequency
RS232	A computer interface used for serial data communications.
RTTY	Radio Teletype. RTTY is a frequency shift digital mode. Characters are sent using alternating Mark and Space tones.
RTTY-L	RTTY transmitted and received on lower sideband (NOR polarity)
RTTY-U	RTTY transmitted and received on upper sideband (NOR polarity)
RTS	'Ready to Send' a com port control line often used for sending the PTT (SEND) command over the CAT interface between the radio and a PC.
RX	Abbreviation for receive or receiver

SDR	Software Defined Radio. The FTDX101 is more correctly a 'Hybrid SDR' as it does not use direct digital sampling except for the spectrum displays.
Sked	A pre-organized or scheduled appointment to communicate with another amateur radio operator
SNR	Signal-to-Noise Ratio in dB (decibels).
Soft Key	A button or selectable icon displayed on the touchscreen
Split	The practice of transmitting on a different frequency to the one that you are receiving on. Split operation is commonly used by DXpeditions and anyone who generates a large pileup of callers. SSB split is commonly 5-10 kHz. CW split is commonly 1-2 kHz.
Squelch	Squelch mutes the audio to the speakers when you are not receiving a wanted signal. When the received signal level increases the squelch opens and you can hear the station. The Squelch setting does affect the audio output over the USB cable.
SSB	Single sideband transmission mode.
SWR	Standing Wave Ratio. The RF power reflected back from a mismatched antenna or connection. (metered)
Tail	Furry attachment at the back of a dog or cat. Also, the length of time a repeater stays transmitting after the input signal has been lost, or a short flexible length of coaxial cable at the antenna or shack end of your main feeder cable.
TU	Morse code abbreviation meaning 'to you.' It is used when you have finished transmitting and wish the other station to respond.
TX	Abbreviation for Transmit or Transmitter
TXW	A button that lets you listen on the transmit frequency during Split operation. Also known as XFC.
UHF	Ultra High Frequency (300 MHz - 3000 MHz).
USB	Universal serial bus – serial data communications between a computer and other devices. USB 2.0 is fast. USB 3.0 is very fast.
USB	Upper sideband SSB transmission.
USOS	Unshift on space. Used to minimize transmission errors on RTTY.
VBW	Video Bandwidth is the ability of the spectrum scope (spectrum and waterfall) to distinguish weak signals from noise. A narrow VBW can filter noise but requires more processing power.
VFO	Variable Frequency Oscillator. The FTDX101 has two VFOs called 'Main' and 'Sub.' The main tuning knob controls the active VFO.
VHF	Very High Frequency (30 MHz -300 MHz)
VOX	Voice Operated Switch. Voice-activated receive to transmit switching
W	Watts – unit of power (electrical or RF).

Table of drawings and images

Figure 1: SSB with compression 14
Figure 2: SSB no compression 14
Figure 3: ALC and Po metering as AMC OUT is adjusted 16
Figure 4: Installed Windows COM ports 36
Figure 5: Renaming the Playback tab 37
Figure 6: Renaming the Recording tab 38
Figure 7: My Playback level FTDX101 TX 41
Figure 8: My Recording level FTDX101 RX 41
Figure 9: N1MM Enhanced COM port setting 43
Figure 10: N1MM Standard COM port setting 43
Figure 11: WSJT COM port settings 45
Figure 12: JS8Call setup 46
Figure 13: MMTTY and EXTFSK 1.06 config 47
Figure 14: MixW 4 has a dropdown selection for the FTDX101 48
Figure 15: MixW 3 settings 49
Figure 16: Fldigi setting for Flrig control 50
Figure 17: Flrig enhanced COM port settings 50
Figure 18: Flrig control software 51
Figure 19: Ham Radio Deluxe setup 52
Figure 20: DM780 PTT setting 53
Figure 21: Colour picking dialogue box 55
Figure 22: Colours 1-11 are for the wideband display, 12-18 are for the narrowband section 55
Figure 23: Adjusting the 2D spectrum and waterfall 59
Figure 24: Adjusting the 3DSS display 59
Figure 25: The VC Tune setting can desensitize the spectrum display 70
Figure 26: Meter display in dual receiver mode 105
Figure 27: Meter display in MONO single receiver mode 105
Figure 28: Metering when the spectrum scope is expanded 106
Figure 29: CW and SSB (with Notch and Contour) Filter Function Displays 106
Figure 30: Dual spectrum and waterfall displays 109
Figure 31: The FUNC menu – orange group 115
Figure 32: The FUNC menu – purple group 116
Figure 33: The FUNC menu – green group 119
Figure 34: The FUNC menu – red group 120
Figure 35: The FUNC menu – blue group 121
Figure 36: The FUNC menu – yellow group 123
Figure 37: FTDX101D rear panel 172
Figure 38: Schematic of Yaesu style keyboard 174

Index

A

AF-OUT 174
AFSK RTTY 28
AFSK RTTY audio levels 41
AFSK RTTY from PC program 28
AGC Soft key 114
ALC 175
Ant 1 173
Ant 2 173
ANT antenna port selection 112
Antenna switching 66
Antenna tuner 78, 156
Antenna tuner jack 176
APF filter and Auto Tune 25
ATT attenuators 112
Audio Codec 36
Author 199

B

Back-up memory channels 62
Back-up radio configuration 63
Band buttons 156
Band switching (Amp or Tuner) 101
Before you update the firmware 64
BPSK and QPSK compared 32
Break-in 78
Break-in setting 24

C

Calibrate TX spectrum scope 187
CAT settings 39, 144
CENTER display 110
Checking installed firmware revision 64
CLAR button 164
Colour choice 56
Colour memories 56
COM Port settings in IC-7610 39
COM Port settings in PC software 39
Computer mouse operation 95
Confusion about AMC 13
Contest number 22, 141
Conventions 2
Cooling fan 173

CS button 82
CURSOR display 110
CW APF filter 84
CW Break-in setting 83
CW key speed and pitch 85
CW memories 141
CW memory (MESSAGE method) 22
CW memory (TEXT method) 21
CW message keyer 23, 25, 85
CW paddle not working properly 181
CW Settings menu 137
CW sidetone 25, 84
CW sidetone but no transmit 187
CW Skimmer 44

D

DC 13.8V 177
Digital Master (DM780) 52, 54
Digital mode on the wrong band 184
DISP button 158
Display your callsign 9
DM780 .... 52, 54
DNF button 167
DNF digital notch filter 72
Driver software 35
DSP filter function displays 106
Dual Spectrum displays 56

E

Elec Key 169
Extension settings menu 153
External antenna tuner 78
External digital mode audio levels 41
External meters 106

F

Fast button 161
FH-2 keypad 174
Firmware update Yaesu website 64
Firmware updates 63
FIX display 111
Fldigi 49
FM split 92
FM-N and FM modes 94

Frequency display 107
Front panel connectors..................... 169
Front panel controls 155
FSK RTTY 29
FT8 preset 33
FUNC button 158
Function menu 115

G

Getting ready for digital modes 33
Glossary 188
GND 177

H

Ham Radio Deluxe 52
Headphone jack 170
Hidden meters 183
HRD 52
Hz digits 107

I

I.F. Out 178

J

JS8 call 46

K

Key 173
Keyer message macros 94
Keyer settings 23
Keyer type 140
Keying speed and pitch 23
kHz digits 107

L

Linear am power - digital modes 61
Linear amplifier 60
Linear amplifier ALC level 61
Linear amplifier connections 60
Lock button 161
Lost receiver audio 183
Lower Spectrum Scope Soft Keys 110

M

M>V button 166
Main AF and RF/Sql 155
Main VFO Knob 163
Main/Sub 163

MAIN/SUB knob 163
MAIN/SUB VFO button 165
Mem Group 144
Memory channels and FM 94
MHz digits 107
Mic Jack 170
MIC scan 146
MixW 48
MMTTY 46
MONI transmit monitor 80
Morse code prosigns 22
Morse keys 23
MOX 79
Multi knob 158

N

N1MM Logger+ 42
NB (Noise Blanker) 166
NOR, REV, RTTY-L, and RTTY-U 30
NR (Digital Noise Reduction) 167

O

OmniRig 44
Onboard RTTY operation 27, 31
Operating CW mode 83
Operating Split 86
Operating Split with a single VFO 89
Operation settings menu 142

P

Panel, Edge or Bar meters 105
Phantom signals on digital modes ... 184
Phone Jack 170
Power button 155
PRESET mode 33
PSK audio levels 32
PSK from an external PC program 32
PSK message keyer 31

Q

QMB (Quick Memory Bank) 144, 164
Quick split input 87, 145

R

R.FIL roofing filter 113
Radio settings for SSB voice 38
Radio settings menu 125
Rear panel connectors 172

Receiver failing to receive 183
REM 174
Repeater offset................................... 92
Repeater Tone 92
Reset.. 154
RTTY Mark and Shift 29
RTTY message keyer 27
RTTY polarity settings........................ 29
RX Out 178

S

S meter stuck................................... 184
S.MENU button 158
SCAN mode... 95
Screen capture 62
Screen saver 180
Screenshot picture 62
SD card 62
SD Card slot 171
Serial numbers 180
Setting the time and date.............9, 153
Setting up a USB connection 34
Setting up PC digital mode software . 41
Setting up the radio............................. 9
Setting up the radio for CW.............. 23
Setting up the radio for PSK 30
Setting up the radio for RTTY 27
Setting up the radio for SSB 10
SO2R single operator two radio 89
SO2V single operator two VFOs 89
Span not following.......................... 183
Spectrum and waterfall display..54, 109
Spectrum Scope controls............57, 187
Spectrum to waterfall display ratio ... 54
Speech processor 12, 81
Split settings........................87, 145, 146
Split without using the SPLIT mode... 89
SSB operation 67
Stop a message being sent 23, 27
Sub AF and RF/Sql 155
SWR meter appears on TX................ 186
SWR meter disappears on TX 186

T

The problem with the FIX display.... 111

Tone Controls................................... 180
Touchscreen calibration................... 154
Touchscreen display functions......... 105
Transceiver stuck in transmit........... 184
Transmitter clarifier 81
Transmitter power overshoot.......... 179
Transmitting....................................... 77
Troubleshooting............................... 179
Tune button 156
Tuner jack 176
Tuning rates 107
TX transmit buttons 80

U

UP and DOWN buttons 146
Updating the firmware 65
Upper spectrum Scope settings 112
USB 1 port (rear panel) 176
USB cable settings DATA-U & DATA-L 38
USB cable settings for the PSK mode. 38
USB cable settings for the RTTY mode
 38
USB Jacks (front panel) 169
USB mouse and keyboard 95, 169

V

V/M button 165
V<M button...................................... 166
VC TUNE buttons 164
VFO focus... 108
Voice memory keyer won't transmit 182
VOX 79

W

Waterfall and 3DSS spectrum colours 55
Waterfall to spectrum display ratio ... 54
Waterfall too dark............................ 180
Waterfall too light............................ 181
WSJT-X software COM settings.......... 44

Z

Zin/SPOT button 157

The Author

Well, if you have managed to get this far you deserve a cup of tea and a chocolate biscuit. It is not easy digesting large chunks of technical information. It is probably better to dip into the book as a technical reference. Anyway, I hope you enjoyed it and that it has made life with the FTDX101 a little easier.

I live in Christchurch, New Zealand. I am married to Carol who is very understanding and tolerant of my obsession with amateur radio. She describes my efforts as "Andrew playing around with radios." We have two children and two cats. James has graduated from Canterbury University with a degree in Commerce and Alex is a doctor working in the Wellington Hospitals.

I am a keen amateur radio operator who enjoys radio contesting, chasing DX, digital modes, and satellite operating. But I am rubbish at sending and receiving Morse code. I write extensively about many aspects of the amateur radio hobby. This is my ninth book.

Thanks for reading my book!

73 de Andrew ZL3DW.

THE END
73 and GD DX

Quick Reference Guide

Function	FUNC menu setting
AGC settings (AM)	<FUNC> <RADIO SETTING> <MODE AM>
AGC settings (CW)	<FUNC> <RADIO SETTING> <MODE CW>
AGC settings (FM)	<FUNC> <RADIO SETTING> <MODE FM>
AGC settings (PSK/DATA)	<FUNC> <RADIO SETTING> <MODE PSK/DATA>
AGC settings (RTTY)	<FUNC> <RADIO SETTING> <MODE RTTY>
AGC settings (SSB)	<FUNC> <RADIO SETTING> <MODE SSB>
AM Mod from USB cable or DATA jack	<FUNC> <RADIO SETTING> <MODE AM> <REAR SELECT>
AM Mod level (Mic)	<FUNC> <RADIO SETTING> <MODE AM> <MIC GAIN>
AM Mod source (Mic or Rear)	<FUNC> <RADIO SETTING> <MODE AM> <AM MOD SOURCE>
AM OUT level to RTTY/DATA jack (Main or Sub)	<FUNC> <RADIO SETTING> <MODE AM> <AM OUT LEVEL>
AM OUT Receiver to RTTY/DATA jack (Main or Sub)	<FUNC> <RADIO SETTING> <MODE AM> <AM OUT SELECT>
AM Power - max RF power on AM	<FUNC> <OPERATION SETTING> <TX GENERAL> <AM MAX POWER>
AMC or Compressor level adjust (leave on COMP)	<FUNC> <OPERATION SETTING> <TX AUDIO> <PROC LEVEL>
ANT3 receiver or transmit antenna settings	<FUNC> <OPERATION SETTING> <GENERAL> <ANT3 SELECT>
APF filter frequency	Main or Sub APF knob
APF filter width	<FUNC> <OPERATION SETTING> <RX DSP> <APF WIDTH>
Audio level from RTTY/DATA port (AM)	<FUNC> <RADIO SETTING> <MODE AM> <RPORT GAIN>
Audio level from RTTY/DATA port (CW)	<FUNC> <RADIO SETTING> <MODE CW> <RPORT GAIN>
Audio level from RTTY/DATA port (FM)	<FUNC> <RADIO SETTING> <MODE FM> <RPORT GAIN>
Audio level from RTTY/DATA port (PSK/DATA)	<FUNC> <RADIO SETTING> <MODE PSK/DATA> <RPORT GAIN>
Audio level from RTTY/DATA port (SSB)	<FUNC> <RADIO SETTING> <MODE SSB> <RPORT GAIN>
Beep level	<FUNC> <OPERATION SETTING> <GENERAL> <BEEP LEVEL>
Beer	<FRIDGE> <OPEN BOTTLE> <DRINK CONTENTS>
BPSK / QPSK switch	<FUNC> <RADIO SETTING> <ENCDEC PSK> <PSK MODE>

Quick reference guide | 201

Break-in delay	<FUNC> <RADIO SETTING> <MODE CW> <CW BREAK_IN DELAY>
Break-in Semi or Full	<FUNC> <RADIO SETTING> <MODE CW> <CW BREAK-IN TYPE>
Callsign on splash screen during boot	<FUNC> <DISPLAY SETTING> <DISPLAY> <MY CALL>
Callsign on splash screen during boot delay	<FUNC> <DISPLAY SETTING> <DISPLAY> <MY CALL TIME>
CAT rate	<FUNC> <OPERATION SETTING> <GENERAL> <CAT RATE>
CAT RTS	<FUNC> <OPERATION SETTING> <GENERAL> <CAT RTS>
Compressor level or AMC adjust (leave on COMP)	<FUNC> <OPERATION SETTING> <TX AUDIO> <PROC LEVEL>
Contour filter depth or boost	<FUNC> <OPERATION SETTING> <RX DSP> <CONTOUR LEVEL>
Contour filter width	<FUNC> <OPERATION SETTING> <RX DSP> <CONTOUR WIDTH>
CS button setting	<FUNC> <OPERATION SETTING> <GENERAL> <CS DIAL>
CS button setting	Press and hold CS button
CW APF filter frequency	Main or Sub APF knob
CW APF filter width	<FUNC> <OPERATION SETTING> <RX DSP> <APF WIDTH>
CW DECODE	<FUNC> <DECODE> - while in CW mode
CW decoder bandwidth	<FUNC> <RADIO SETTING> <KEYER> <CW DECODE BW>
CW filter display	<FUNC> <RADIO SETTING> <MODE CW> <CW FREQ DISPLAY>
CW indicator on filter display	<FUNC> <RADIO SETTING> <MODE CW> <CW INDICATOR>
CW initial contest number	<FUNC> <RADIO SETTING> <KEYER> <CONTEST NUMBER>
CW Messages	<FUNC> <REC/PLAY> - while in CW mode
CW messages - auto repeat delay	<FUNC> <RADIO SETTING> <KEYER> <REPEAT INTERVAL>
CW messages - input via Text or Paddle	<FUNC> <RADIO SETTING> <KEYER> <CW MEMORY 1-5>
CW Mod from USB cable or DATA jack	<FUNC> <RADIO SETTING> <MODE CW> <REAR SELECT>
CW Mod source (Mic or Rear)	<FUNC> <RADIO SETTING> <MODE CW> <CW MOD SOURCE>
CW Number Style	<FUNC> <RADIO SETTING> <KEYER> <NUMBER STYLE>
CW OUT level to RTTY/DATA jack (Main or Sub)	<FUNC> <RADIO SETTING> <MODE CW> <CW OUT LEVEL>

CW OUT Receiver to RTTY/DATA jack (Main or Sub)	<FUNC> <RADIO SETTING> <MODE CW> <CW OUT SELECT>
CW sidetone level	in CW mode <FUNC> <MONI LEVEL> <turn MULTI>
CW wave shape	<FUNC> <RADIO SETTING> <MODE CW> <CW WAVE SHAPE>
CW while in SSB mode	<FUNC> <RADIO SETTING> <MODE CW>
Date & Time setting	<FUNC> <EXTENSION SETTING> <DATE & TIME>
Decode from main receiver	<FUNC> <OPERATION SETTING> <GENERAL> <DECODE RX SELECT>
Decode from sub receiver	<FUNC> <OPERATION SETTING> <GENERAL> <DECODE RX SELECT>
DNR level	<FUNC> <OPERATION SETTING> <RX DSP> <DNR LEVEL>
DNR level	<FUNC> <DNR LEVEL> <turn MULTI knob>
Emergency frequency enable	<FUNC> <OPERATION SETTING> <TX GENERAL> <EMERGENCY FREQ TX>
External display - 800x480 or 800x600	<FUNC> <DISPLAY SETTING> <EXT MONITOR> <PIXEL>
External display - enable	<FUNC> <DISPLAY SETTING> <EXT MONITOR> <EXT DISPLAY>
External tuner - set which antenna	<FUNC> <OPERATION SETTING> <GENERAL> <TUNER SELECT>
Filter slope (AM)	<FUNC> <RADIO SETTING> <MODE AM>
Filter slope (CW)	<FUNC> <RADIO SETTING> <MODE CW>
Filter slope (FM)	<FUNC> <RADIO SETTING> <MODE FM>
Filter slope (PSK/DATA)	<FUNC> <RADIO SETTING> <MODE PSK/DATA>
Filter slope (RTTY)	<FUNC> <RADIO SETTING> <MODE RTTY>
Filter slope (SSB)	<FUNC> <RADIO SETTING> <MODE SSB>
Firmware - current firmware versions	<FUNC> <EXTENSION SETTING> <SOFT VERSION>
Firmware update	<FUNC> <EXTENSION SETTING> <SD CARD> <FIRMWARE UPDATE> <DONE>
FM Mod from USB cable or DATA jack	<FUNC> <RADIO SETTING> <MODE FM> <REAR SELECT>
FM Mod source (Mic or Rear)	<FUNC> <RADIO SETTING> <MODE FM> <FM MOD SOURCE>
FM OUT level to RTTY/DATA jack (Main or Sub)	<FUNC> <RADIO SETTING> <MODE FM> <FM OUT LEVEL>
FM OUT Receiver to RTTY/DATA jack (Main or Sub)	<FUNC> <RADIO SETTING> <MODE FM> <FM OUT SELECT>
Headphone main/sub mix	<FUNC> <OPERATION SETTING> <GENERAL> <HEADPHONE MIX>
Keyboard language	<FUNC> <OPERATION SETTING> <GENERAL> <KEYBOARD LANGUAGE>

Keyed data shift	<FUNC> <RADIO SETTING> <MODE PSK/DATA> <DATA SHIFT (SSB)>
LED brightness	<FUNC> <DISPLAY SETTING> <DISPLAY> <LED DIMMER>
Manual notch filter frequency	Main or Sub NOTCH knob
Manual notch filter width	<FUNC> <OPERATION SETTING> <RX DSP> <IF NOTCH WIDTH>
Memory groups (on or off)	<FUNC> <OPERATION SETTING> <GENERAL> <MEM GROUP>
Mouse pointer speed	<FUNC> <DISPLAY SETTING> <DISPLAY> <MOUSE POINTER SPEED>
MPVD ring encoder steps per revolution	<FUNC> <OPERATION SETTING> <TUNING> <MPVD STEPS PER REV>
NB noise blanker level	<FUNC> <NB LEVEL> <turn MULTI knob>
NB noise blanker rejection (attenuation applied)	<FUNC> <OPERATION SETTING> <GENERAL>
NB noise blanker width	<FUNC> <OPERATION SETTING> <GENERAL>
Parametric equalizer SSB (transmitter)	<FUNC> <OPERATION SETTING> <TX AUDIO> <PRMTRC EQ...>
Parametric equalizer with speech compressor turned on	<FUNC> <OPERATION SETTING> <TX AUDIO> <P PRMTRC EQ...>
PC keying line for CW	<FUNC> <RADIO SETTING> <MODE CW> <PC KEYING>
Power - max RF power on 6m band	<FUNC> <OPERATION SETTING> <TX GENERAL> <50M MAX POWER>
Power - max RF power on 70 MHz band	<FUNC> <OPERATION SETTING> <TX GENERAL> <70M MAX POWER>
Power - max RF power on AM	<FUNC> <OPERATION SETTING> <TX GENERAL> <AM MAX POWER>
Power - max RF power on HF bands	<FUNC> <OPERATION SETTING> <TX GENERAL> <HF MAX POWER>
PSK DECODE	<FUNC> <DECODE> - while in PSK mode
PSK Messages	<FUNC> <REC/PLAY> - while in PSK mode
PSK tone offset	<FUNC> <RADIO SETTING> <MODE PSK/DATA> <PSK TONE>
PSK transmit level	<FUNC> <RADIO SETTING> <ENCDEC PSK> <PSK TX LEVEL>
PSK/DATA Mod from USB cable or DATA jack	<FUNC> <RADIO SETTING> <MODE PSK/DATA> <REAR SELECT>
PSK/DATA Mod source (Mic or Rear)	<FUNC> <RADIO SETTING> <MODE PSK/DATA> <PSK/DATA MOD SOURCE>
PSK/DATA OUT level to RTTY/DATA jack (Main or Sub)	<FUNC> <RADIO SETTING> <MODE PSK/DATA> <PSK/DATA OUT LEVEL>
PSK/DATA OUT Receiver to RTTY/DATA jack (Main or Sub)	<FUNC> <RADIO SETTING> <MODE PSK/DATA> <PSK/DATA OUT SELECT>

QMB 5 or 10 memory slots	<FUNC> <OPERATION SETTING> <GENERAL> <QMB CH>
QPSK Polarity Settings	<FUNC> <RADIO SETTING> <ENCDEC PSK> <QPSK POLARITY (RX or TX)>
QSK delay	<FUNC> <RADIO SETTING> <MODE CW> <QSK DELAY TIME>
Quick split input popup enable	<FUNC> <OPERATION SETTING> <GENERAL> <QUICK SPLIT INPUT>
Quick split offset	<FUNC> <OPERATION SETTING> <GENERAL> <QUICK SPLIT FREQ>
Recall FUNC MENU settings from SD card	<FUNC> <EXTENSION SETTING> <SD CARD> <MENU LOAD> <DONE> <Select a file>
Recall memory channels from SD card	<FUNC> <EXTENSION SETTING> <SD CARD> <MEM LIST LOAD> <DONE> <Select a file>
Repeater offset 10m band	<FUNC> <RADIO SETTING> <MODE FM> <RPT SHIFT (28MHz)>
Repeater offset 6m band	<FUNC> <RADIO SETTING> <MODE FM> <RPT SHIFT (50MHz)>
Reset - Full reset	<FUNC> <EXTENSION SETTING> <RESET> <ALL RESET>
Reset - Memory slots	<FUNC> <EXTENSION SETTING> <RESET> <MEMORY CLEAR>
Reset - Menu settings	<FUNC> <EXTENSION SETTING> <RESET> <MENU CLEAR>
RF Gain or Squelch control	<FUNC> <OPERATION SETTING> <GENERAL> <RF/SQL VR>
RF Power setting	<FUNC> <RF POWER> <turn MULTI>
RS232 rate & timeout	<FUNC> <OPERATION SETTING> <GENERAL> <232C....>
RTS/DTR/DAKY selection (AM)	<FUNC> <RADIO SETTING> <MODE AM> <RPPT SELECT>
RTS/DTR/DAKY selection (CW)	<FUNC> <RADIO SETTING> <MODE CW> <RPPT SELECT>
RTS/DTR/DAKY selection (FM)	<FUNC> <RADIO SETTING> <MODE FM> <RPPT SELECT>
RTS/DTR/DAKY selection (PSK/DATA)	<FUNC> <RADIO SETTING> <MODE PSK/DATA> <RPPT SELECT>
RTS/DTR/DAKY selection (RTTY)	<FUNC> <RADIO SETTING> <MODE RTTY> <RPPT SELECT>
RTS/DTR/DAKY selection (SSB)	<FUNC> <RADIO SETTING> <MODE SSB> <RPPT SELECT>
RTTY DECODE	<FUNC> <DECODE> - while in RTTY mode
RTTY Mark Frequency	<FUNC> <RADIO SETTING> <MODE RTTY> <MARK FREQUENCY>
RTTY Messages	<FUNC> <REC/PLAY> - while in RTTY mode
RTTY OUT level to RTTY/DATA jack (Main or Sub)	<FUNC> <RADIO SETTING> <MODE RTTY> <RTTY OUT LEVEL>

RTTY OUT Receiver to RTTY/DATA jack (Main or Sub)	<FUNC> <RADIO SETTING> <MODE RTTY> <RTTY OUT SELECT>
RTTY Polarity (RX)	<FUNC> <RADIO SETTING> <MODE RTTY> <POLARITY RX>
RTTY Polarity (TX)	<FUNC> <RADIO SETTING> <MODE RTTY> <POLARITY TX>
RTTY Shift	<FUNC> <RADIO SETTING> <MODE RTTY> <SHIFT FREQUENCY>
RTTY text and Baudot code settings	<FUNC> <RADIO SETTING> <ENCDEC RTTY>
Save FUNC MENU settings to SD card	<FUNC> <EXTENSION SETTING> <SD CARD> <MENU SAVE> <DONE> <NEW> <ENT>
Save memory channels to SD card	<FUNC> <EXTENSION SETTING> <SD CARD> <MENU MEM LIST> <DONE> <NEW> <ENT>
Scan - Mic scan resume time	<FUNC> <OPERATION SETTING> <GENERAL> <MIC SCAN RESUME>
Scan started from top Mic buttons	<FUNC> <OPERATION SETTING> <GENERAL> <MIC SCAN>
Screensaver	<FUNC> <DISPLAY SETTING> <DISPLAY> <SCREEN SAVER>
SD Card - capacity and free space information	<FUNC> <EXTENSION SETTING> <SD CARD> <INFORMATIONS> <DONE>
SD Card - format	<FUNC> <EXTENSION SETTING> <SD CARD> <FORMAT> <DONE>
SD card - recall FUNC menu settings	<FUNC> <EXTENSION SETTING> <SD CARD> <MENU LOAD> <DONE> <Select a file>
SD Card - recall memory channels	<FUNC> <EXTENSION SETTING> <SD CARD> <MEM LIST LOAD> <DONE> <Select a file>
SD card - save FUNC menu settings	<FUNC> <EXTENSION SETTING> <SD CARD> <MENU SAVE> <DONE> <NEW> <ENT>
SD Card - save memory channels	<FUNC> <EXTENSION SETTING> <SD CARD> <MENU MEM LIST> <DONE> <NEW> <ENT>
Spectrum scope 2D display sensitivity	<FUNC> <DISPLAY SETTING> <SCOPE> <2D DISP SENSITIVITY>
Spectrum scope 3DSS display sensitivity	<FUNC> <DISPLAY SETTING> <SCOPE> <3DSS DISP SENSITIVITY>
Spectrum scope center or carrier point display	<FUNC> <DISPLAY SETTING> <SCOPE> <SCOPE CTR>
Spectrum scope resolution bandwidth	<FUNC> <DISPLAY SETTING> <SCOPE> <RBW>
Split - Quick split input popup enable	<FUNC> <OPERATION SETTING> <GENERAL> <QUICK SPLIT INPUT>
Split - Quick split offset	<FUNC> <OPERATION SETTING> <GENERAL> <QUICK SPLIT FREQ>
Squelch or RF Gain control	<FUNC> <OPERATION SETTING> <GENERAL> <RF/SQL VR>

SSB Mod from USB cable or DATA jack	<FUNC> <RADIO SETTING> <MODE SSB> <REAR SELECT>
SSB Mod source (Mic or Rear)	<FUNC> <RADIO SETTING> <MODE SSB> <SSB MOD SOURCE>
SSB OUT level to RTTY/DATA jack (Main or Sub)	<FUNC> <RADIO SETTING> <MODE SSB> <SSB OUT LEVEL>
SSB OUT Receiver to RTTY/DATA jack (Main or Sub)	<FUNC> <RADIO SETTING> <MODE SSB> <SSB OUT SELECT>
Touchscreen brightness	<FUNC> <DISPLAY SETTING> <DISPLAY> <TFT DIMMER>
Touchscreen calibration	<FUNC> <EXTENSION SETTING> <CALIBRATION>
Touchscreen contrast	<FUNC> <DISPLAY SETTING> <DISPLAY> <TFT CONTRAST>
Transmit bandwidth (AM)	<FUNC> <RADIO SETTING> <MODE AM> <TX BPF SEL>
Transmit bandwidth (CW)	<FUNC> <RADIO SETTING> <MODE CW> <TX BPF SEL>
Transmit bandwidth (FM)	<FUNC> <RADIO SETTING> <MODE FM> <TX BPF SEL>
Transmit bandwidth (PSK/DATA)	<FUNC> <RADIO SETTING> <MODE PSK/DATA> <TX BPF SEL>
Transmit bandwidth (SSB)	<FUNC> <RADIO SETTING> <MODE SSB> <TX BPF SEL>
Transmit monitor level	in a voice mode <FUNC> <MONI LEVEL> <turn MULTI>
Transmit time-out timer	<FUNC> <OPERATION SETTING> <GENERAL> <TX TIME OUT TIMER>
Tuning step dial rate AM, FM, other modes	<FUNC> <OPERATION SETTING> <TUNING> <AM, FM, CH STEP>
VFO display font	<FUNC> <DISPLAY SETTING> <DISPLAY> <FREQ STYLE>
VFO encoder steps per revolution	<FUNC> <OPERATION SETTING> <TUNING> <MAIN STEPS PER REV>
VFO step size RTTY & PSK 5 Hz or 10 Hz	<FUNC> <OPERATION SETTING> <TUNING> <RTTY/PSK DIAL STEP>
VFO step size SSB & CW 5 Hz or 10 Hz	<FUNC> <OPERATION SETTING> <TUNING> <SSB/CW DIAL STEP>
Voice Messages	<FUNC> <REC/PLAY> - while in SSB, AM, or FM mode
VOX level from DATA jack	<FUNC> <OPERATION SETTING> <TX GENERAL> <DATA VOX GAIN>

Made in United States
Troutdale, OR
01/05/2025

27652440R00117